CTBUH Technical Guide

Outrigger Design for High-Rise Buildings

An output of the CTBUH Outrigger Working Group

Hi Sun Choi, Goman Ho, Leonard Joseph & Neville Mathias

Council on Tall Buildings and Urban Habitat

ILLINOIS INSTITUTE OF TECHNOLOGY

Routledge
Taylor & Francis Group

NEW YORK AND LONDON

Bibliographic Reference:
Choi, H., Ho, G., Joseph, L. & Mathias, N. (2012) *Outrigger Design for High-Rise Buildings: An output of the CTBUH Outrigger Working Group.* Council on Tall Buildings and Urban Habitat: Chicago.

Principal Authors: Hi Sun Choi, Goman Ho, Leonard Joseph & Neville Mathias
Coordinating Editor & Design: Steven Henry
Layout: Tansri Muliani

First published 2012 by the Council on Tall Buildings and Urban Habitat

This edition published 2014 by Routledge
2 Park Square, Milton Park, Abingdon, Oxon, OX14 4RN

Simultaneously published in the USA and Canada by Routledge
711 Third Avenue, New York, NY 10017

Routledge is an imprint of the Taylor & Francis Group, an informa business

Published in conjunction with the Council on Tall Buildings and Urban Habitat (CTBUH) and the Illinois Institute of Technology

© 2012 Council on Tall Buildings and Urban Habitat

ISBN13 978-0-939493-34-0

Council on Tall Buildings and Urban Habitat
S.R. Crown Hall
Illinois Institute of Technology
3360 South State Street
Chicago, IL 60616
Phone: +1 (312) 567-3487
Fax: +1 (312) 567-3820
Email: info@ctbuh.org
http://www.ctbuh.org

Front Cover: Shanghai Tower, China (see pages 64–65) © Gensler

Principal Authors

Hi Sun Choi, Thornton Tomasetti, Inc.
Goman Ho, Arup
Leonard Joseph, Thornton Tomasetti, Inc.
Neville Mathias, Skidmore, Owings & Merrill, LLP

Peer Review Panel

Ahmad Abdelrazaq, Samsung C&T Corporation
Charles Besjak, Skidmore, Owings & Merrill LLP
Ryan Chung, Dongyang Structural Engineers
Paul Fu, Thornton Tomasetti, Inc.
Ramon Gilsanz, Gilsanz Murray Steficek LLP
Andrew Hakin, WSP Group
Yasuyoshi Hitomi, Nihon Sekkei, Inc.
Ronald Johnson, Skidmore, Owings & Merrill LLP
Sang Dae Kim, Korea University
Ron Klemencic, Magnusson Klemencic Associates
Cori Kwitkin, Thornton Tomasetti, Inc.
Peng Liu, Arup
Larry Novak, Portland Cement Association
Juneid Qureshi, Meinhardt Pte., Ltd.
Mark Sarkisian, Skidmore, Owings & Merrill LLP
David Scott, Laing O'Rourke
Simon Shim, Thornton Tomasetti, Inc.
Robert Sinn, Thornton Tomasetti, Inc.
Fei-fei Sun, Tongji University
Paul Tsang, Arup
David Vesey, Arup
Masayuki Yamanaka, Obayashi Corp.

Contents

About the CTBUH

The Council on Tall Buildings and Urban Habitat is the world's leading resource for professionals focused on the design and construction of tall buildings and future cities. A not-for-profit organization based at the Illinois Institute of Technology, Chicago, the group facilitates the exchange of the latest knowledge available on tall buildings around the world through events, publications, research, working groups, web resources, and its extensive network of international representatives. Its free database on tall buildings, The Skyscraper Center, is updated daily with detailed information, data, images, and news. The CTBUH also developed the international standards for measuring tall building height and is recognized as the arbiter for bestowing such designations as "The World's Tallest Building."

About the Authors

Hi Sun Choi
Thornton Tomasetti, Inc.

Hi Sun Choi is a Senior Principal at Thornton Tomasetti and has approximately 20 years of experience in structural analysis, investigation, design, and review of a variety of building types, including commercial and residential buildings. Her expertise includes the design of supertall buildings for seismic risk assessment, building motion due to wind, performance-based design, and waterfront developments on reclaimed land.

Goman Ho
Arup Hong Kong Ltd.

Dr. Goman Ho is a Director at Arup. He joined Arup in 1992 after his postgraduate study. He has been significantly involved in a large number of tall building and long span projects, from analysis and design to construction. His research interests include stability and nonlinear transient analysis. He is the past-president of the ASCE Hong Kong, current fellow member of the HKISC, and editor of the International Journal of Advanced Steel Construction.

Leonard Joseph
Thornton Tomasetti, Inc.

With more than 35 years of experience, Leonard Joseph has analyzed, designed, and reviewed high-rise buildings, sports facilities, hangars, hotels, historic buildings, manufacturing facilities, and parking garages. He works with a wide variety of materials, including structural steel, reinforced concrete, precast and post-tensioned concrete, masonry, wood, and light gage framing. For buildings around the world, Len deals with seismic, wind, and other environmental hazards, and incorporates local construction practices into his designs.

Neville Mathias
Skidmore, Owings & Merrill LLP

Neville Mathias is an Associate Director and Senior Structural Engineer with Skidmore, Owings & Merrill, LLP. He has worked extensively on the structural design of major buildings across California and around the world for the past 26 years. He specializes in the seismic design of non-prescriptive buildings using performance based, non-linear methodologies.

Preface

Outrigger systems have come into widespread use in supertall buildings constructed since the 1980s, eclipsing the tubular frame systems previously favored. Their popularity derives largely from the unique combination of architectural flexibility and structural efficiency that they offer, compared to tubular systems with characteristic closely spaced columns and deep spandrel girders. Despite extensive recent use, outrigger systems are not listed as seismic load resisting systems in current building codes, and specific design guidelines for them are not available. Recognizing the pressing need for such guidelines, the CTBUH formed the Outrigger Working Group, launched in September 2011, charged with developing a design guide.

Objectives of this Guide

This design guide provides an overview of outrigger systems including historical background, pertinent design considerations, design recommendations, and contemporary examples. The guide has three objectives for serving the engineering profession. First, by gaining familiarity with the unique considerations surrounding outrigger systems, designers will be better prepared to determine if outriggers are appropriate for use in a given situation. Second, if designers choose to apply an outrigger system, the guide provides technical background information necessary to understand and address key issues associated with outrigger system use. Examples also illustrate the broad range of solutions applied to these issues, since outrigger designs are not typically "one size fits all." The third objective supports this point by presenting key issues and recommendations; the guide provides a framework for further discussions within the industry. Rather than being the "last word" in outrigger system design, future editions of the guide should reflect expanded and revised information.

> **Outrigger system performance is affected by outrigger locations through the height of a building, the number of levels of outriggers provided, their plan locations, the presence of belt trusses to engage adjacent perimeter columns versus stand alone mega columns, outrigger truss depths, and the primary structural materials used.**

Content Overview

Outrigger systems function by tying together two structural systems – typically a core system and a perimeter system – to yield whole-building structural behaviors that are much better than those of the component systems. They do this by creating a positive interaction between the two tied systems. The beneficial effect is most pronounced where the responses of the component systems under lateral loads are most disparate. Outriggers find excellent use, for example, in tall buildings that utilize dual lateral systems including a perimeter frame. The very different cantilever type deformations of core structures and the portal type deformation of frame structures under lateral loads are harnessed to best effect at a given level to maximize the benefit of outrigger systems in these structures. Outriggers also prove beneficial when engaging perimeter columns that would otherwise be gravity-only elements. In contrast, outriggers are less effective for "tube in tube" dual systems because core and perimeter tubes exhibit similar cantilever deformation behaviors even before they are linked.

Outrigger system performance is affected by outrigger locations through the height of a building, the number of levels of outriggers provided, their plan locations, the presence of belt trusses to engage adjacent perimeter columns versus stand alone mega columns, outrigger truss depths, and the primary structural materials used.

Tying together core and perimeter structural systems with outriggers creates unique design and construction problems to resolve. Most significantly, particularly in concrete and mixed-material structures, different levels of axial stress and strain in core and perimeter vertical members cause differential shortening which increases over time due to creep and shrinkage. Differential movement can cause enormous forces in outrigger members attempting to tie the two systems together. "Virtual" outrigger systems eliminate direct outriggers connecting core and perimeter systems by instead using belt trusses in combination with stiff and strong diaphragms. Although less effective than direct outriggers, "virtual" outriggers have been developed and used to overcome the challenges posed by differential shortening, along with other benefits. Additional solutions to address the issue of differential shortening have been developed and implemented, including shimming and construction sequencing approaches, and the very innovative use of damping mechanisms to address slow, long term movements and provide opportunities for enhanced structural damping without impacting fundamental outrigger action.

These and a host of other relevant topics have been addressed in this guide, including capacity design approaches, connection design, thermal effects, and more. The apparent conflict of outrigger systems with traditional seismic code requirements are discussed, such as story stiffness and story strength ratio requirements as well as strong column-weak beam requirements. For example, outrigger systems add strength and stiffness beyond what is normally available to specific locations over a structure's height but stiffness and strength ratio requirements in codes are meant to guard against sudden reductions in the normal values of these quantities; not increases. Similarly, strong column-weak beam requirements developed to protect against story mechanisms in frame structures have less relevance where the core provides a large percentage of available story shear strength. The applicability of traditional code requirements such as these at outrigger floors thus needs careful consideration of structural first principles and discussion with building officials and peer reviewers prior to incorporation.

The Outrigger Working Group hopes this guide is useful to design professionals and code writers, and looks forward to receiving feedback which will be used to improve future editions.

1.0 Introduction to Outrigger Systems

1.0 Introduction to Outrigger Systems

1.1 Background

Outriggers are rigid horizontal structures designed to improve building overturning stiffness and strength by connecting the building core or spine to distant columns. Outriggers have been used in tall, narrow buildings for nearly half a century, but the design principle has been used for millennia. The oldest "outriggers" are horizontal beams connecting the main canoe-shaped hulls of Polynesian oceangoing boats to outer stabilizing floats or "amas" (see Figure 1.1). A rustic contemporary version of this vessel type illustrates key points about building outrigger systems:

▶ A narrow boat hull can capsize or overturn when tossed by unexpected waves, but a small amount of *ama* flotation (upward resistance) or weight (downward resistance) acting through outrigger leverage is sufficient to avoid overturning. In the same manner, building outriggers connected to perimeter columns capable of resisting upward and downward

forces can greatly improve the building's overturning resistance.

▶ Even though a boat may be ballasted to resist overturning it can still experience uncomfortable long-period roll, outrigger-connected *amas* greatly reduce that behavior and shorten the period of the movement. Similarly, building outriggers can greatly reduce overall lateral drift, story drifts, and building periods.

▶ Boats can have outriggers and *amas* on both sides or on one side. Buildings can have a centrally located core with outriggers extending to both sides or a core located on one side of the building with outriggers extending to building columns on the opposite side.

The explanation of building outrigger behavior is simple: because outriggers act as stiff arms engaging outer columns, when a central core tries to tilt, its rotation at the outrigger level induces a tension-compression couple

in the outer columns acting in opposition to that movement. The result is a type of restoring moment acting on the core at that level.

Analysis and design of a complete core-and-outrigger system is not that simple: distribution of forces between the core and the outrigger system depends on the relative stiffness of each element. One cannot arbitrarily assign overturning forces to the core and the outrigger columns. However, it is certain that bringing perimeter structural elements together with the core as one lateral load resisting system will reduce core overturning moment, but not core horizontal story shear forces (see Figures 1.2 & 1.3). In fact, shear in the core can actually increase (and change direction) at outrigger stories due to the outrigger horizontal force couples acting on it.

Belts, such as trusses or walls encircling the building, add further complexity. Belts can improve lateral system efficiency. For towers with outriggers engaging individual mega column, belts can direct more gravity load to the mega columns to minimize net uplift, reinforcement or the column splices required to resist tension and stiffness reduction associated with concrete in net tension. For towers with external tube systems – closely spaced perimeter columns linked by spandrel beams – belts reduce the shear lags effect of the external tube, more effectively engage axial stiffness contributions of multiple columns, and more evenly distribute across multiple columns the large vertical forces applied by outriggers. For both mega column and tube buildings, belts can further enhance overall building stiffness through virtual or indirect outrigger behavior provided by high in-plane shear stiffness (discussed later), as well as increasing tower torsional stiffness. Belts working with mega columns can also create a

▲ Figure 1.1: Samoan outrigger canoe. © Teinesavaii.

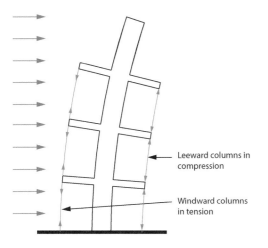

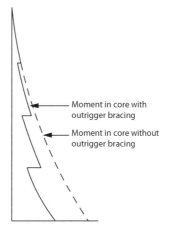

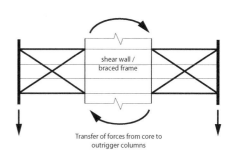

Leeward columns in compression

Windward columns in tension

Moment in core with outrigger bracing

Moment in core without outrigger bracing

shear wall / braced frame

Transfer of forces from core to outrigger columns

▲ Figure 1.2: Interaction of core and outriggers. (Source: Taranath 1998)

▲ Figure 1.3: Outrigger at core. (Source: Nair 1998)

secondary lateral load resisting system, in seismic engineering terminology.

A core-and-outrigger system is frequently selected for the lateral load resisting system of tall or slender buildings where overturning moment is large compared to shear, and where overall building flexural deformations are major contributors to lateral deflections such as story drift. In such situations, outriggers reduce building drift and core wind moments. Because of the increased stiffness they provide, outrigger systems are very efficient and cost-effective solutions to reduce building accelerations, which improves occupant comfort during high winds (Po & Siahaan 2001).

1.2 Benefits of an Outrigger System

Deformation Reduction
In a building with a central core braced frame or shear walls, an outrigger system engages perimeter columns to efficiently reduce building deformations from overturning moments and the resulting lateral displacements at upper floors. A tall building structure which incorporates an outrigger system can experience a reduction

in core overturning moment up to 40% compared to a free cantilever, as well as a significant reduction in drift depending on the relative rigidities of the core and the outrigger system (Lame 2008). For supertall towers with perimeter mega columns sized for drift control, reduction in core overturning can be up to 60%. The system works by applying forces on the core that partially counteract rotations from overturning. These forces are provided by perimeter columns and delivered to the core through direct outrigger trusses or walls, or indirect or "virtual" outrigger action from belt trusses and diaphragms as described in Section 3.6.

Efficiency
For systems with belt trusses that engage all perimeter columns, columns already sized for gravity load may be capable of resisting outrigger forces with minimal changes in size or reinforcement, as different load factors apply to design combinations with and without lateral loads. In the event that additional overall flexural stiffness is required, the greater lever arm at outrigger columns makes additional material more effective than in the core. Outriggers may also permit optimization of the overall building system using techniques such

as the unit load method to identify the best locations for additional material (Wada 1990). By significantly decreasing the fraction of building overturning moment that must be resisted by the core, wall, or column material quantities in the core can be reduced while outrigger, perimeter belt, and column quantities are increased by a smaller amount. Lower limits on core required strength and stiffness may be defined by story shears resisted by the core alone between outrigger levels, special loading conditions that exist at outrigger stories, or short-term capacity and stability if outrigger connections are delayed during construction

Foundation Forces
A separate but related advantage is force reduction at core foundations. Outrigger systems help to effectively distribute overturning loads on foundations. Even where a foundation mat is extended over the full tower footprint, a core-only lateral system applying large local forces from overturning can generate such large mat shear and flexural demands, as well as net tension in piles or loss of bearing, that the design becomes uneconomical or impractical. Reducing core overturning and involving perimeter column axial

forces to help resist overturning from lateral loads reduces mat shear demand, flexural demand, and net uplift conditions by spreading loads from overturning across the tower footprint. Reducing variations in sub-grade stresses or pile loads under the core from lateral load will reduce foundation rotations that can contribute to overall and inter-story drifts. Having an outrigger system may or may not change other aspects of the foundation design, such as governing pile loads and footing or mat bearing pressures. They must be checked for all relevant load combinations, as combinations for gravity loads may govern over combinations including lateral loads.

Gravity Force Transfers
Outriggers and belt trusses can help reduce differential vertical shortening between columns, or between a column and the core. This can reduce floor slopes between those elements which may occur from creep, shrinkage, or thermal changes. The reduction is achieved by force transfers between adjacent columns through belt trusses, or between the columns and core through outriggers. This is a secondary benefit at best, and is a two-edged

Outriggers are rigid horizontal structures designed to improve building overturning stiffness and strength by connecting the building core or spine to distant columns.

sword: force transfers can become quite large – potentially comparable in magnitude to forces from overturning resistance – and costly to achieve. Balancing potential benefits and costs requires a solid understanding of the phenomenon as well as proper application of details and construction strategies to manage its effects. Force determination and control is discussed later in the text.

Torsional Stiffness
Belt trusses can provide a different secondary benefit: improved torsional stiffness. A core-only tower can have low torsional stiffness compared to a perimeter-framed tower, due to the much smaller distance between resisting elements. A core-and-outrigger building can have similarly low torsional stiffness. Belt trusses can force perimeter columns to act as fibers of a perimeter tube that, while not as stiff as a continuous framed tube, still provides significant additional torsional stiffness.

Disproportionate Collapse Resistance
Another potential benefit related to force transfer capability is disproportionate (progressive) collapse resistance. On projects which require considering sudden loss of local member or connection capacity, outriggers can provide alternate load paths. For example, where perimeter columns are engaged by belt trusses, loads from floors above a failed perimeter column could "hang" from the upper column acting in tension and then be transferred through upper belt trusses to adjacent undamaged columns. Where outriggers are present without belt trusses, it may be possible to hang upper floor loads from outriggers which load the core, but massive outrigger columns may be too heavily loaded for this load path to be practical. In a braced-frame-core building, loads from floors above a failed core column could be shared by

perimeter members through outriggers. Of course the design must be checked to confirm that alternate load paths can accept the resulting forces rather than leading to further failures. For disproportionate collapse checks load factors are often smaller and capacities considered are often larger than those used for the basic design, so the effect of these conditions on the building design may be minimal, depending on the scenarios considered.

For example, 300 Madison Avenue, New York City (2003) is of moderate height (35 stories / 163 meters) but includes belt trusses at floors 9 through 11 and above 35 as indirect or virtual outriggers to reduce overturning on the slender core, improve torsional stiffness, and provide alternate load paths in case of perimeter column damage (Arbitrio & Chen 2005; Chen & Axmann 2003).

Architectural Flexibility
Core-and-outrigger systems permit design variations in exterior column spacing to satisfy aesthetic goals and, in some cases, specific functional requirements. Internal or direct outriggers need not affect the building's perimeter framing or appearance compared with other floors. Supertall buildings with outriggers may have a few exterior mega columns on each face, which opens up the façade system for flexible aesthetic and architectural expression. This overcomes a primary disadvantage of closed-form tubular systems used in tall buildings. The quantity and location of mega columns have impacts on typical floor framing, plans featuring widely-separated columns and column-free corners may require deep and heavy spandrels for the strength, deflection control, and vibration control requirements of long spans and cantilevers. The core-and-outrigger approach is scalable, with potential applicability to buildings 150 stories tall or more (Ali & Kyoung 2007).

1.3 Challenges for Outrigger System Design

Incorporating an outrigger system in the design of a tall building is not a trivial exercise. The conditions below must be considered and resolved for a successful core-and-outrigger design. Where the designer has control over them, recommended approaches for these conditions are provided later in the document.

Usability of Occupied Spaces

Because outrigger systems include elements in vertical planes (walls, truss diagonals) they can potentially interfere with occupiable or rentable space. However, this drawback can be diminished through architectural and structural planning. One common strategy is to locate outriggers at floors with major mechanical spaces, or at refuge areas required by local codes or practices. Major mechanical levels are often double-height spaces, which is advantageous for deeper outrigger trusses with more efficient chords and diagonals than shallower trusses. Using mechanical floors requires careful coordination with mechanical room layouts, access requirements, and service routes to avoid potential conflicts. Equipment must be located far enough from trusses or walls to permit maintenance. Ducts, pipes, and conduit banks must cross the outrigger planes using dedicated openings. Because the resulting mechanical layouts may be less efficient than usual, more space may be needed for mechanical rooms than normally anticipated.

Outrigger Story Locations

There are ideal locations for outriggers, but the realities of space planning to suit architectural, mechanical, and leasing criteria usually make such considerations purely academic. Outrigger locations are typically limited to mechanical or refuge floors.

▲ Figure 1.4: One Liberty Place, Philadelphia.
© Rainer Vertlbock

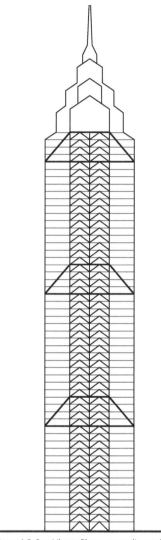

▲ Figure 1.5: One Liberty Place – superdiagonal system. © Thornton Tomasetti

Even so, acceptable performance can usually be achieved using these locations. Where mechanical levels do not occur at appropriate elevations, a "superdiagonal" strategy has been used, most notably at One Liberty Place in Philadelphia (see Figure 1.4 & 1.5), where sets of four-story-tall diagonals run through occupied space. This arrangement qualifies as an outrigger system rather than a full-building-width braced system because the diagonals occur only at widely separated locations. By running diagonals through four stories, the obstructed width on any floor is less than one-quarter of the clear floor span. Tenants conceal these braces behind cabinets and partitions with minimal impact to floor functionality. However, any outrigger placement strategy affecting occupied spaces must consider its likely acceptability in the local leasing environment.

Diaphragm Forces, Stiffness, and Details

Floor diaphragms interact with outrigger systems in multiple ways. In a direct or conventional outrigger system, models with incorrect or unrealistic diaphragm properties will report incorrect force values in outrigger chords supporting slabs, as well as incorrect building deformations. Diaphragm stiffness modeling is particularly important for indirect "virtual" outrigger/belt truss systems, as the diaphragms are key elements in the load paths that make the system work. Overly optimistic diaphragm stiffness will overestimate outrigger participation and underestimate building drift and core overturning forces. Too-low diaphragm stiffness assumptions will underestimate the forces experienced by the diaphragms, belt trusses, and perimeter columns. In both direct conventional and indirect "virtual" outrigger/belt truss systems, diaphragm modeling parameters can affect shear and bending deformations and the resulting forces in perimeter columns and framing that braces them, as columns follow the building slope changes above and below outrigger levels. Designs should envelope reasonable ranges for diaphragm stiffness. Where diaphragms alone do not offer sufficient stiffness for effective virtual outrigger/belt truss performance, horizontal bracing beneath the floor slab can be provided. The presence of under-floor horizontal bracing can affect material quantities, construction time, and coordination with multiple trades potentially impacted by the additional members.

Differential Vertical Shortening

Conventional or directly-framed outrigger system designs must address the potential for load redistribution between columns and core resulting from differential axial strains. All buildings experience differential vertical shortening between core and perimeter vertical members acting at different stresses under gravity load. An all-steel system with a braced core will experience predictable shortening that

is virtually complete once the building is occupied. Buildings that include concrete columns or walls will additionally experience long-term vertical deformations due to cumulative creep and shrinkage strains. The magnitude and timing of such deformations will differ between members as stresses, concrete mixtures, volume-to-surface ratios, and reinforcing ratios. This makes prediction of differential movements a complex time and sequence-based challenge.

Outriggers connecting a core wall or core columns to perimeter columns must be designed for possible force transfers from differential shortening that occurs between the two connected vertical members after the connection date. If not specifically mitigated, the transfer forces can become very large, in some cases approaching the design forces from wind. Since designing members and connections for large additional anticipated forces is a considerable penalty, minimizing the effect through proportioning (relative stiffness of the outrigger system), construction sequencing, special detailing, or other methods is a worthwhile effort. For example, final bolting of steel outrigger diagonal connections at the Jin Mao Building in Shanghai was postponed until the end of construction as described in the example in Section 3.5 as was final concreting of Vierendeel outrigger frame posts at the Petronas Towers in Kuala Lumpur. Shims and jacks at outrigger tip bearing connections of Cheung Kong Centre and Two International Finance Centre permit shim removal and replacement as needed to have outriggers function during the construction period and still minimize force transfer over time. The projects are discussed further in Sections 3.4 and 3.5 respectively. Another strategy is to simply avoid direct connections between columns and the core. That would not avoid

sloping or warping of floors from differential shortening, but it would avoid force transfers. This is a key benefit of the indirect or virtual outrigger/belt truss approach discussed later in the text.

Differential Thermal Strains

Force transfers can also occur through conventional or direct outriggers when the columns and core experience differential temperature conditions, as from perimeter columns exposed to weather. Forces in conventional or direct outriggers from differential temperatures can be significant where columns are fully exposed, as at the New York Times Building (see Section 3.2) (Scarangello et al. 2008), but this is not a common condition.

Foundation Dishing

Foundation loads are concentrated under the central core of a tall building, both because that element supports a large fraction of the tower floor area, and because a concrete core (if present) represents a large fraction of the total building dead load. Even if loads were evenly distributed across a large foundation, settlement is typically greater under its center than at its edges, because of the way that stresses and strains spread throughout the sub-grade. Load concentration and sub-grade behavior act together to cause foundation dishing, causing the core to displace downward relative to the perimeter columns. This is a different phenomenon than differential shortening but the result is the same: if conventional or direct outriggers connect the core and perimeter columns, dishing can induce force transfers between vertical elements. The relative magnitude and timing of this phenomenon will differ from creep and shrinkage, and their interaction and effect on outriggers will vary by project. In an all-concrete structure where the core shortens less than perimeter columns, dishing will reduce the differential shortening effect. In a concrete core and steel-framed perimeter system, dishing will add to the differential shortening effect.

Connection Forces and Details

Outriggers located at just a few points along building height tend to generate large forces based on relative stiffness. For these large forces to help counteract core overturning moments, they must be transmitted from columns to outrigger trusses or walls, and then to the core. When the same material is used in all members the magnitude of forces requires large, often geometrically complex connections. All-steel systems may use large bolted gusset plates, field-welded joints, or a mix of approaches at different locations. All-concrete systems require sufficient room in all members to pass and develop reinforcing bars while permitting effective concrete placement. The challenge increases at mixed systems such as steel outrigger trusses between concrete mega columns and concrete core walls, and at composite systems with steel members embedded within or enclosing concrete. Possible approaches for steel truss connection to concrete or composite members include, but are not limited to, pockets, flush embedded plates, embedded steel cores, embedded steel stubs, and embedded "anchor trusses." The most appropriate solution will depend on the forces involved, the materials available, the space available, erection equipment available, and local construction preferences. Detailed discussions and examples are provided in Section 2.8.

Construction Schedule

The presence of numerous special members and heavy connections of outriggers and belt trusses, along with the changes from typical floor framing at outrigger levels, can significantly slow down the erection process. Schedule impact can be minimized by developing an optimized erection schedule and clear erection guidelines (Ali & Kyoung 2007). Various strategies for constructing outriggers have been developed based on local environments and construction conditions. For example, the South China region that includes Hong Kong experiences an average of five strong storms and typhoons per year, making concrete core and core-and-outrigger lateral systems popular for economically providing occupant comfort through lateral stiffness and inherent damping. Delaying outrigger connections to allow for initial core shortening would reduce gravity load transfer forces, but outriggers must respond to potentially high winds during the construction period, and for deflection control for partial operation of buildings still under construction, a common local practice. This has led to development of several creative solutions to outrigger construction which are explained in detail in Section 2.10.

Seismic Design Criteria

Outrigger system seismic design is not discussed in building codes, unlike many other systems and combinations of systems. For example in the China code outriggers are classified as "strengthened floors" with specifications on the floors but no explicit requirements on outrigger trusses, and outrigger performance levels are normally determined in a Seismic Expert Review meeting. In the ASCE 7-05 Standard (ASCE 2005) referenced in model building codes such as the International Building Code, there are 82 systems and combinations, none of which address outriggers. This omission is not surprising since no single standard design approach is suitable for all outrigger situations. Outriggers and belt trusses are stiff and strong elements at discrete locations within a structure. This can be inconsistent with seismic design approaches based on distributed stiffness and strength. Strong outriggers may also apply forces large enough to load other elements to the point of damage and non-ductile behavior. Seismic design approaches successfully used for outrigger systems include performance based design (PBD), in which suites of scaled seismic time histories and nonlinear building models are used to demonstrate that the building performs well, and capacity based design, in which outrigger members with predictable, ductile behavior are used to avoid overloading other building elements. These approaches are further discussed in Section 2.13.

Change in Story Stiffness

Soft-story seismic provisions in model building codes typically look at the change in story stiffness from one story to the next story up. In an outrigger system the outrigger floors exhibit smaller story drift from a reverse shear force in the core. As a result, outriggers could be considered as inherently "stiff-stories" and the stories immediately below an outrigger are always "soft-stories." Some researchers have raised the issue and recommended minimizing outrigger stiffness in high seismic regions (Cheng et al.1998). However, there are both semantic and practical resolutions to this issue. If story stiffness, or force/story drift, is based on core shear force rather than story shear force, calculated stiffness is not that different from other stories. Regardless of story stiffness definitions, tall buildings will have many typical or "soft" stories between outrigger levels which can provide distributed ductile behavior over most of the building height. This is very different from the code writers' concern about drift being concentrated at just a few "soft-stories." The alternative, intentionally softening outriggers and stiffening perimeter columns to maintain more uniform story stiffness is theoretically possible, but probably impractical in most situations. Outriggers soft enough to avoid the stiffness "jump" may not be stiff enough to provide effective reductions in drift and core overturning moment. Weak-story seismic provisions can be similarly tricky to navigate.

Strong Column Weak Beam Provision
Strong column seismic provisions in model building codes are intended to avoid the undesirable phenomenon of all columns in a story yielding and developing hinges at top and bottom, potentially leading to story collapse. Requiring column flexural strength to be greater than beam strength at each joint results in columns capable of acting as continuous spines, encouraging beam yielding that is well distributed over the height of the structural moment frame. In a core-and-outrigger system, the strong column weak beam provision does not appear necessary or appropriate at perimeter columns because the central core walls or core braced bays already provide a strong spine. The strong column weak beam philosophy could be appropriately applied to the interaction of outriggers and the core through capacity based design limiting outrigger forces, or performance based design evaluating forces from realistic seismic excitation. This is discussed further in Section 2.14.

1.4 Conditions Less Suitable for Outrigger Systems

Shear Deformations
Outriggers efficiently reduce core overturning forces and resulting building deformations. Structural systems governed by story shear deformations, such as moment frames, would not benefit enough from outriggers to justify their cost.

Core Flexural Stiffness
Outrigger systems interact with cores based on relative stiffness. If a core is already comparatively stiff, as described by a low aspect ratio (building height/core width), it may be impractical to provide further stiffness through outriggers since the outrigger and column member sizes to accomplish this may be much larger than strength alone would dictate. For this reason, the height at which outriggers start to be efficient is often higher for office towers with wide cores typically containing washrooms, mechanical rooms, numerous elevators, and several stairwells; and lower for residential towers with narrow, slender cores containing minimal elevators and stairs. To improve core efficiency, some residential designs have cores enlarged by enclosing some rooms of adjacent residential units.

Lack of Symmetry
Outriggers are most effective when symmetrically distributed about a central core since this provides the largest distance for the force couple between outrigger columns, maximizing the benefit they provide. It also relieves core overturning moment without imposing additional net axial loads on the core. An unsymmetrical system may have outrigger force couples involving axial forces in the core, complicating its analysis and design. Force transfers due to differential shortening will cause a symmetrical system to deform straight downward. Force transfers in an unsymmetrical system can result in an overturning moment cranked into the building, leading to lateral displacements under gravity loads. However, this does not mean unsymmetrical systems cannot be used. There are examples of successful unsymmetrical outrigger systems that address the above concerns in design.

Torsional Concerns
Conventional outrigger systems help reduce core overturning forces and overturning-related deformations. A floor plan with core located to one side may be susceptible to torsional deformations and torsion-induced forces affecting the design. If controlling torsional forces and deformations is of primary importance, a perimeter tube (frame) or belt truss system would be more effective than an outrigger system without belts.

Differing Material Properties
Force transfer from differential shortening is a fact of life for conventional, direct outriggers. It can be well managed, using methods described later in this text, when similar materials are used in the core and the perimeter columns. Steel shortening is elastic and well defined. Concrete long-term shortening is time-dependent and larger, but with concrete columns and a concrete core the difference is of interest, and that is a small fraction of the total shortening values. It may not be practical to manage the effect when different materials are used in the core and the perimeter columns. The difference between steel and concrete shortening behavior becomes quite large over time, and the difference grows after construction is complete, so any post-construction mitigation would require call-back work. Clever solutions for continuous adjustment based on linked hydraulic systems have been proposed (Kwok & Vesey 1997) as described later in this document, but placing long-term reliance on such devices for structural elements as key as outriggers is debatable at best. Conventional, direct outriggers have been used in mixed construction, typically concrete cores and steel columns, only after detailed study of likely effects and mitigation measures. Indirect virtual outriggers/belt trusses offer a way to sidestep the force transfer issue, as discussed later.

Column Size Limitations
Outrigger effectiveness often requires the ability to adjust column stiffness.

This may mean increasing column sizes, especially for outriggers located high in a building, since increased stiffness from a larger area offsets the softening effect of the column length. If column sizes are strictly limited it may restrict their usefulness in an outrigger system.

Impact on Other Trades

Both conventional direct outriggers and indirect virtual outriggers/belt trusses affect space usability. If a mechanical floor design is already tight, it may not be able to accommodate the space occupied by outriggers. Or mechanical and refuge floors may not occur at elevations for efficient outriggers, such as being located only at low floors. In such situations the disruptions caused by an outrigger system may not be worthwhile for the project overall.

1.5 Types of Outrigger Systems

A direct or conventional outrigger system consists of a core with shear stiffness, outrigger columns along the building perimeter, and stiff outrigger trusses or walls oriented in vertical planes and projecting horizontally from core to perimeter. Lateral loads causing overturning moment and rotation of the core at outrigger levels will try to move outrigger truss tips up and down. Outrigger columns restrain the movement and generate opposing forces. Those forces crank the outriggers in the opposite direction, inducing a reverse

story shear in the core that reduces overturning moments and rotations (see Figures 1.2 & 1.3 in Section 1.1).

An indirect or "virtual" outrigger or belt truss system provides similar behaviors without direct connections to core walls in the vertical plane (see Figure 1.6). Instead, a belt truss completely rings the building perimeter and engages perimeter columns. Lateral loads causing overturning moment and rotation of the core at belt truss levels will try to move floor diaphragms on different floors left and right. The belt truss engaging both floors tries to follow and rotate itself by moving one face up and one face down. Perimeter columns restrain the movements and generate opposing forces. Typically the corner columns develop the greatest forces. Those vertical forces act through the belt truss to create horizontal forces in the floor diaphragms in the opposite direction, inducing a reverse story shear in core that reduces overturning moments and rotations.

The process of movement, restraint, opposing forces, and overturning reduction hinges on the relative stiffness of the core and of the outrigger-and-column system. The load path for conventional outriggers is more direct so it provides restraint more efficiently than "virtual" outriggers. But in some circumstances the indirect or virtual outrigger approach is sufficient to meet the needs of a tall building, and

it avoids the complex connections of outrigger truss ends.

Core-and-outrigger systems have been constructed using steel, concrete, composite (steel members encased within concrete or filled with concrete), and mixed members (separate steel and concrete members). One common structural system with conventional or direct outriggers has a concrete core with projecting concrete outrigger walls or steel outrigger trusses that engage perimeter columns. Cores of steel braced frames and steel plate shear walls would typically engage steel outrigger trusses. The outrigger columns at the far end of outrigger trusses may be structural steel, reinforced concrete, composite concrete-encased steel members, or composite concrete-filled

> **Outriggers are most effective when symmetrically distributed about a central core since this provides the largest distance for the force couple between outrigger columns, maximizing the benefit they provide. It also relieves core overturning moment without imposing additional net axial loads on the core.**

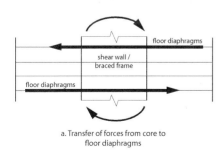

a. Transfer of forces from core to floor diaphragms

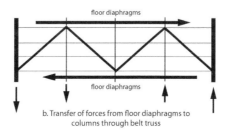

b. Transfer of forces from floor diaphragms to columns through belt truss

▲ Figure 1.6: Virtual outrigger or belt wall system. (Source: Nair 1998)

steel boxes or pipes. Material selections for the core, outriggers, and columns depend on multiple factors including required design strength, required stiffness, space limitations, connection forces and details, material availability, construction methodology, and schedule. The selection process should also consider that using different materials in the core and the outrigger columns can lead to large values of differential shortening capable of generating large transfer forces or requiring elaborate mitigation measures.

1.6 Historical Outrigger Systems in Buildings

The widespread popularity of tall building outrigger systems today can be seen as a response to fundamental disadvantages of the perimeter and bundled tube frame systems developed in the 1960s and 1970s in the United States. Tall buildings such as the World Trade Center twin towers in New York and Sears Tower and the John Hancock Center in Chicago had relatively dense exterior frames with closely spaced columns and deep spandrel beams. In fact the majority of lateral load resistance was provided by the exterior frame, with little or no contribution from the building core. While they were without question structurally efficient from a material utilization standpoint, such systems had a strong presence on the building exterior which ultimately proved too limiting for architectural aesthetic freedom; the structural frame itself often determined the overall visual composition of the tower.

Properly proportioned core-and-outrigger schemes offer far more perimeter flexibility and openness for tall buildings than the perimeter moment or braced frames and bundled tubes that preceded them. Spandrel beams sized for gravity loads alone can

▲ Figure 1.7: Tour de la Bourse, Montreal. © Jeangagnon

be relatively shallow. Column spacing can be adjusted to match architectural requirements. The system also allows for significant discontinuities in exterior form, including notches and setbacks. Compared to the direct structural expression of perimeter tube frame buildings, core-and-outrigger towers, with a few exceptions, have tended to reveal very little of their underlying structural logic.

A very early example of an outrigger structure is the 47-story Tour de la Bourse (formerly Place Victoria Tower) in Montreal (see Figure 1.7 & 1.8), designed by Nervi and Moretti and built in 1965. Prior to the Place Victoria project, essentially all high-rise buildings were built with steel frames. This building is 190 meters high and was the first concrete structure to integrate outriggers. Nervi's structural design followed the principal of using fewer but larger columns to concentrate dead load, so they would perform as compression members at all times, regardless of external forces. The primary structural system consists of a core, large corner columns, and four levels of X-braced transverse outrigger trusses connecting the core to these columns. Two intermediate columns on each side of the tower make up the secondary system and serve to support the floor structure.

The core-and-outrigger system was extended to steel-framed towers by Fazlur Khan for the 41-story,152-meter tall BHP House, now 140 William Street, in Melbourne, Australia (see Figure 1.9) completed in 1972; and in the 42-story, 183-meter tall, 120,000 square meter First Wisconsin Center, now U.S. Bank Center (see Figure 1.10), built in 1974 in Milwaukee, Wisconsin. Both of the office towers are similar in having three belt trusses, located at the bottom, middle, and top, with the middle and top trusses acting to distribute forces from direct outrigger trusses to adjacent perimeter columns (Khan 2004). The bottom belt provides for a change in column spacing. Mechanical equipment levels coincide with the outrigger levels.

Core-and-outrigger designs have evolved since these pioneering projects. The greatest changes probably resulted from the development of high strength concrete, which provided both strength and stiffness economically in compression elements such as core walls and columns. Building examples later in this document illustrate the current relationship of materials and structural systems through a wide variety of projects and locations.

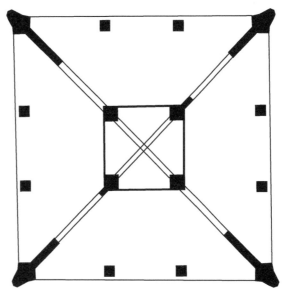

Layout of the tower at the mechanical floors showing the connection of the core to the corner column. Note the eight lateral columns remain independent of the primary structural system.

▲ Figure 1.8: Tour de la Bourse outrigger layout in plan. (Source: Thornton Tomasetti)

▲ Figure 1.9: 140 William Street, Melbourne. © Property Council of Australia

▲ Figure 1.10: U.S. Bank Center, Milwaukee. © Marshall Gerometta/CTBUH

2.0 Design Considerations for Outrigger Systems

2.0 Design Considerations for Outrigger Systems

2.1 Appropriate Conditions for Outrigger Systems

All multi-story buildings require at least one core to accommodate elevators, stairs, mechanical shafts, and other common services. Because views are a significant part of the intrinsic value in tall buildings, it is most common for their core or cores to be centrally located within the floor plan to place occupants along exterior walls. A central core also locates the center of lateral stiffness close to the center of lateral wind load and center of mass for lateral seismic loads, minimizing torsional forces. In high-seismic regions many tall buildings have a dual system, sometimes called "core and frame" or "tube in tube," with a perimeter moment frame providing significant torsional stiffness but a smaller contribution to overturning stiffness. When the core is relatively large in plan it may be wide enough to provide strength against overturning and stiffness against drift. However, a core becomes less efficient as the height/core width aspect ratio increases. For an aspect ratio exceeding eight or so, the structural premium to control drift and resist overturning is

When direct or conventional outrigger walls or trusses are not acceptable for the building due to space limitations, an indirect, "virtual" outrigger or belt truss system may be used.

large enough to consider introducing outriggers. The building height for which this occurs is typically lower for residential buildings with small cores for isolated stairwells and elevator shafts than for office buildings with larger cores including washrooms and mechanical rooms. Some residential tower designs include cores enlarged by enclosing occupied rooms as well as elevator banks for this reason. For constant core properties, drift from flexural or overturning behavior will increase approximately as the cube of building height (Lame 2008). To maintain the building drift/height ratio below a particular criterion, as building height doubles, core stiffness would have to quadruple. But simply thickening core walls for more stiffness would reduce rentable area. Introducing outriggers can alleviate the dependence on the core system and maximize useful space between the core and exterior columns.

When direct or conventional outrigger walls or trusses are not acceptable for the building due to space limitations or a column layout which is not aligned with the core walls, an indirect, "virtual" outrigger or belt truss system may be used. Behavior of the exterior columns is tied to behavior of the core through stiff belts and strong, stiff floor diaphragms at upper and lower levels of each belt. This approach eliminates complicated outrigger connections at columns and at the core. It minimizes concerns about inadvertent load transfers between core and perimeter from differential shortening. Alternatively, a belt truss can be used together with direct, conventional outriggers to engage more, smaller columns rather than requiring fewer, larger mega columns. This results in more uniform perimeter column sizes where desired.

By engaging more perimeter columns through the outrigger system, the structure will gain more stiffness, transfer

more of the overturning moment from core to perimeter, and better distribute overturning forces across the foundation. The belt truss is most efficient when a belt wraps around the entire perimeter of the building and engages all exterior columns. It is recommended that the gravity system be optimized together with the lateral system from an early design phase so that the outrigger column design can be at maximum efficiency – putting column area where it can do the most good, and designing belts with consideration of gravity load transfers among columns through the belts. The system selected has a large influence on the design approach. In a mega column design gravity load is intentionally concentrated only at those columns connected to direct outriggers. The mega columns receive floor loads either by long span floor framing or by pickup trusses spanning between mega columns, with the pickup trusses interrupting and supporting secondary columns. This way full dead load is available to offset uplift forces, and column material needed for gravity strength also provides helpful axial stiffness. In a direct outrigger building design with numerous columns on each face, it may be necessary to increase stiffness of the belt truss diagonals and the perimeter columns to each side of the outrigger column to equalize distribution of the outrigger force among the columns and receive maximum benefit from their stiffness. In an indirect or virtual outrigger belt truss design, corner columns tend to provide most overturning resistance, but may not attract much of the gravity load unless specific attention is paid to relative stiffness of all system elements. Ideally the same member sizes work for strength and for stiffness. But requiring a truss to redistribute loads may result in increased material quantities to satisfy strength requirements.

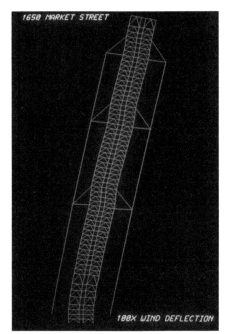

Figure 2.1: One Liberty Place deflected shape.
© Thornton Tomasetti

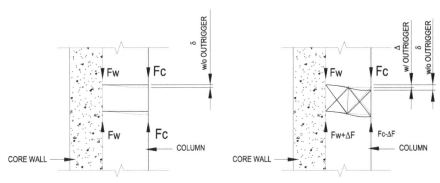

▲ Figure 2.2: Transfer of gravity loads from columns to core. © Thornton Tomasetti

2.2 Load Transfer Paths in Outrigger Systems

When a structure containing an outrigger system is loaded laterally, the outriggers resist core rotation by using perimeter columns to push and pull in opposition, introducing a change in the slope of the vertical deflection curve as seen in Figure 2.1, a portion of the core overturning moment is transferred to the outriggers and, in turn, tension in windward columns and compression in leeward columns (see Figure 1.2 in Section 1.1). Typically gravity load on columns is sufficient to maintain net compression, but net tension must always be checked, starting just below the topmost outriggers. At concrete columns net tension could result in cracking and dramatic, if temporary, reduction in axial stiffness that affects system behavior.

The magnitude of drift reduction and core overturning moment from each outrigger level is a function of

several building properties: core flexural stiffness, outrigger flexural and shear stiffness, outrigger locations along the building height, plan dimensions between core-and-outrigger centroids, and axial stiffness of the outrigger columns (Lame 2008). Depending on relative stiffness of the core and the outrigger system, core story shear could reverse to the point of being greater in absolute value than the story shears above and below.

The same stiff outriggers that generate interaction between core and columns under lateral loads will also cause interaction under vertical loads. Differential shortening, whether from elastic shortening, inelastic creep, and shrinkage or thermal effects will lead to forces being transferred between core and columns through the outriggers. Because it is more likely that columns will be acting at higher stress than core walls under gravity loads, outriggers typically tend to transfer outer column gravity load to the core when core and columns are of the same material (see Figure 2.2). With a concrete core and steel perimeter columns, the effect reverses over time as creep and shrinkage causes the core to shorten more. Load transfer effects can be minimized through control of construction sequence or use of special connection details to be discussed later in the text.

When an indirect or virtual outrigger, sometimes called a belt truss, is used, no outrigger walls or trusses directly connect the core and columns. Core rotation tilt cannot cause outrigger trusses to push and pull on perimeter columns. Instead, core rotation is used to horizontally displace stiff floor diaphragms connected to the top and bottom chords of a belt truss. As the horizontal displacements try to rotate the belt, it resists that rotation through vertical push and pull force couples in the outrigger columns (see Figure 1.6 in Section 1.5), with axial forces concentrated at building corners due to the shear lag effect. More uniform participation of all perimeter columns occurs only when belt trusses are combined with direct outriggers. Virtual outrigger behavior only occurs at a large change in perimeter lateral stiffness, as at a change from moment frame to trussed story. Belt trusses can also help equalize gravity loads among perimeter columns.

Horizontal diaphragm forces enter the belt wall system through shear studs on belt truss chords, or a concrete-to-concrete connection using reinforcing steel for a concrete structure. Realistic estimates of diaphragm shear stiffness, including stiffness reductions from shear and tension cracking, are important for predicting indirect or virtual outrigger behavior and their contribution to overall building behavior, as discussed later in the text. Diaphragm shear behavior may be viewed as a group of shear panels bounded by chord and drag elements, or may be viewed using strut-and-tie methodology, with compression struts extending diagonally from core corners to points on the perimeter, where forces are resolved by belt chords and drags in tension. For either case, realistic diaphragm stiffness is much smaller than gross dimensions and solid concrete properties would indicate.

One major advantage of the belt wall system is that it is not affected by differential inelastic vertical deformations between core and perimeter, so no vertical load transfer occurs between the core wall and perimeter columns. However, a belt truss can experience vertical load transfer forces if it tries to equalize axial strains that differ between adjacent perimeter columns.

In regions of high seismicity using belt walls for both lateral load resistance as indirect (virtual) outriggers and gravity load transfers for interrupted columns may be problematic. Belt wall or belt truss member yielding in a large earthquake could unacceptably reduce or alter the gravity load paths. Solutions may include protecting the belt walls from this situation by limiting capacity elsewhere along the lateral load paths, or avoiding belt member yielding through a performance based design process.

2.3 Determining Locations of Outriggers in Elevation

The degree to which an outrigger system provides improvement of building stiffening and reduction of building drift depends in part on the number and locations of outriggers. Outrigger locations and effectiveness are driven by four issues.

Number of Outrigger Sets
More outrigger sets provide more opportunities for rotation restraint which leads to drift reduction. However, each additional outrigger set comes with costs: it takes additional time and effort for erection, and it typically interrupts work flow compared to that at typical floors. Impact on the overall building schedule can be minimized by applying special construction strategies such as the wall blockouts used at Two International Finance Centre, Hong Kong as described in Section 3.5. Even if the total material quantity is unchanged between two designs, distributing it across more outriggers means more pieces to erect. On the other hand, relying on fewer outriggers to minimize piece counts and involve fewer non-typical floors may result in members so heavy that they require higher-capacity, more costly erection equipment, etc. Costs and benefits must be weighed. Furthermore, as discussed later under "Space Availability," the possible locations for outriggers are usually dictated by the program of a building. Outriggers are usually located in mechanical floors and refuge floors only.

Direct or Indirect (Virtual) Outriggers
As the names indicate, the shorter load path from column to core by direct outriggers makes them stiffer and more efficient. To achieve the same stiffness benefit, indirect outriggers (belt trusses or walls) would be required on more floors than direct outriggers. This trade off is rarely an issue in reality, the particular benefits of each outrigger type lead to their use in different building conditions. Both outrigger types can also be present in the same building, as where multiple outriggers offer desired stiffness and strength benefits, but not every outrigger level desired can accommodate direct outrigger trusses, or where differential shortening is more problematic for direct outriggers at some levels than at others.

Spacing to Equalize Distances From Outriggers To Core Inflection Points
Figure 2.3 illustrates this point using drastically simplified examples, each with a single lateral force at top, uniform core flexural properties, infinite core shear stiffness, and infinite outrigger properties. For the same number of outriggers, changing their placement in these examples can change roof drift by more than 50%. Of course the decisions are not so clear-cut for actual building designs with distributed lateral loads, varying core properties, realistic outrigger truss and column stiffness, and design criteria different from top floor drift.

Outrigger Column and Truss Stiffness
To develop and apply forces countering core overturning, outrigger trusses and outrigger columns must be stiff as well as strong. Column axial stiffness is more easily achieved when the vertical distance from foundation to outrigger is shorter, even though a lower outrigger level may not be optimal in theory. Optimization methods such as the unit load method (Wada 1990) can help identify the relative importance of different members to outrigger system stiffness for meeting a particular displacement goal. Overall system stiffness can also be improved by greatly increasing member sizes locally in acceptable areas rather than along the length of an element. Examples include outrigger columns oversized at basement and lobby floors, but not at office floors where leasable area is critical, and floor diaphragms of indirect or virtual outrigger/belt truss systems being much thicker than typical floor slabs, if headroom permits. However, there is a limit to this strategy: the columns in an outrigger system are

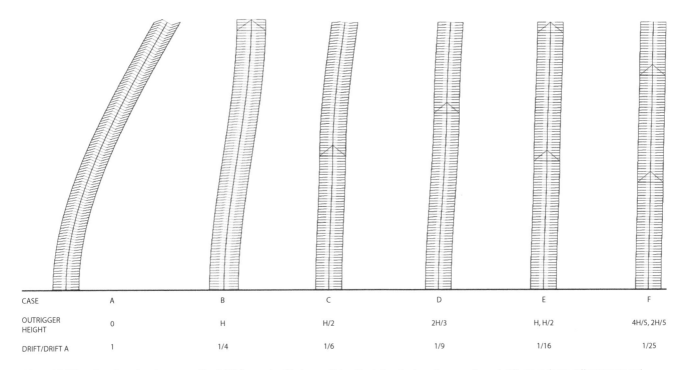

CASE	A	B	C	D	E	F
OUTRIGGER HEIGHT	0	H	H/2	2H/3	H, H/2	4H/5, 2H/5
DRIFT/DRIFT A	1	1/4	1/6	1/9	1/16	1/25

▲ Figure 2.3: Effect of outrigger locations on roof level drift from a simplified case of lateral load at roof only, uniform core flexural stiffness, and very stiff outriggers and outrigger columns. © Thornton Tomasetti

effectively story-high springs in series, so compensating for low stiffness in some locations can require an impractically large quantity increase in other locations. Another concern is that stiffness may vary with load, if uplift forces applied to outrigger columns by outrigger trusses or walls may exceed minimum dead load compression at upper stories. Steel column designs can address this directly by providing column splices with significant tensile capacity. At reinforced concrete or composite columns, the embedded steel must be capable of resisting the net tension and the reduced axial stiffness of a column cracked in tension must be considered in lateral analyses. Since the effect is nonlinear, reflecting it can require multiple iterations of analyses or use of nonlinear software.

Space Availability

For most buildings this issue dominates all the others. Contemporary building designs usually (but not always) include a major mechanical space at or near the building top, making that a natural location for an outrigger, even if a single top outrigger is not optimal compared to a single outrigger at perhaps 2/3 height due to core flexure, column stiffness, and net tension considerations. Additional outriggers are typically associated with intermediate mechanical floors as determined by the selected mechanical services design, or with refuge floors where required by local practice. Such opportunities may occur every 12 to 25 stories (Gerasimidis et al. 2009). Where opportunities for outriggers are closely spaced, which levels to use for outriggers, if any, should be determined by the other issues mentioned previously in this section.

Studies (Gerasimidis et al. 2009) of optimal outrigger placement show that, assuming a top outrigger will be present at a top mechanical floor, a second outrigger of equal stiffness would ideally be located at building mid-height to control overall drift. If the second outrigger is not of equal stiffness, its optimum location for drift control may differ. If the second outrigger location is established by other criteria such as space availability, its stiffness should be tuned to maximize efficiency. Tuning could involve adjusting member sizes for outrigger trusses, or column areas below and above the second outrigger. A further complication is that outrigger systems are indeterminate: outrigger stiffness at any one level is based on both the contribution of trusses at that level, and of the columns engaged, with outriggers at different levels typically engaging the same perimeter columns. As a result, the optimal arrangement of outrigger locations and member sizes to minimize lateral load responses will largely depend on the pattern of column size changes with height.

With the understanding that any guidance "rules of thumb" are based on sets of assumptions regarding core, column and outrigger stiffness, lateral load distribution, and the parameter of interest; studies show that the optimum location for a single outrigger is at one-quarter to two-thirds of building height, measured up from ground level (see Figure 2.3). Having such a wide range of potentially optimum heights illustrates the complexity inherent in outrigger placement. For example, an outrigger as low as one-quarter of height would seem too low to provide much overturning relief to a core, but it has the advantage of shorter, and therefore stiffer, outrigger columns. Another general guideline, for optimum performance of a structure with "n" outrigger levels, states outriggers should be placed at the $1/(n+1)$ up to the $n/(n+1)$ height locations (Smith & Coull 2007; Bayati et al. 2008). For one outrigger the guideline would indicate a location at half of the building height. For two outriggers, one-third and two-third height are optimal. However, if one of the outriggers must be at the top the second truss would optimally be at half height; others have suggested the second outrigger location should be at 60% of building height. If there are three outriggers; one-quarter, one-half, and three-quarter height points are optimal, but if one is a top outrigger the others should be at one-third and two-third height. As discussed above, any selection of outrigger locations must consider both the realities of space availability and the influence of member size decisions on this indeterminate system.

2.4 Diaphragm Floors

Understanding diaphragm behavior is important for any outrigger system. If a belt wall or virtual outrigger system is used, a stiff floor diaphragm is required at the top and bottom chord of each belt wall in order to transfer the core bending moment, in the form of floor shear and axial forces, to the belt wall and eventually to the columns. Diaphragm stiffness and strength is essential to the successful performance of belt wall systems. Indeed, the floors at belt walls are significantly thicker, or specially trussed, to provide that stiffness and strength. However the effect must not be exaggerated: a simple rigid-diaphragm modeling assumption must not be used. Improperly modeled diaphragms will result in misleading behaviors and load paths, and incorrect member design forces, for both indirect "virtual" outrigger/belt truss systems and direct, conventional outrigger systems.

Modeling Indirect Outrigger Floors

To analyze stresses and investigate performance of diaphragm floors in an indirect or virtual outrigger/belt truss system, a three dimensional finite element model of the outrigger system including the core wall, belt wall, columns, and flexible diaphragm floor is strongly recommended. All significant floor openings should be reflected in the model to determine and design for stress concentrations in the diaphragm which may exist around these openings. Such a model can also help determine appropriate in-plane load paths required for resolution of potentially large forces, such as those discussed for the strut and tie analogy. In simple situation a carefully design 2D slab model can provide necessary slab design information, but it will rely on forces determined from other models or require making assumptions that could affect results. Three dimensional FEM detail can be incorporated in the overall building model, or can be determined through a sub-model whose properties are then reflected in simplified fashion within the overall model. For example, dummy diagonal floor bracing in an overall model, avoiding rigid diaphragms at or near the outrigger levels, could provide core-to-belt load paths and stiffness values determined in a sub-model. Recent software analysis features such as semi-rigid diaphragms could also be studied for use.

Modeling Direct Outrigger Floors

If outriggers are present to connect the core wall and columns, a portion of core overturning moment can be transferred to the columns directly through the outrigger. It would seem that the diaphragm is not of interest in this case. However, outrigger systems and floor slabs still interact at outriggers and at outrigger columns. Where outrigger trusses or walls have slabs connected to chord members or counted as flanges, slab behavior will affect outrigger behavior. Treating floor slabs as rigid diaphragms in analytical structural models is common and computationally efficient, but a rigid diaphragm or master/slave node approach should not be used at or near outrigger floors as it will artificially stiffen the outrigger system, erroneously report zero force in outrigger truss chords, and obscure forces needed to determine force resolution and compatibility with the floor diaphragm. Instead, modeling should consider a range of possible slab stiffness contributions. Find deflections and forces with the slab disengaged from the member, or with the area of local slab reinforcement treated as additional outrigger chord area. Then consider varying degrees of slab participation, up to the stiffness of an uncracked slab tributary to the chord, as these could control for maximum outrigger shear or diagonal forces determined by relative stiffness. Also consider crack control reinforcing for the slab based on induced strains found in this study.

Away from the outrigger trusses themselves, diaphragms participate in force redistribution from outriggers among parallel core walls. Outriggers that engage some, but not all, core wall or braced bay lines deliver restoring forces locally that must then be distributed across all core elements. Modeling floors as rigid diaphragms or through master/slave nodes result in unrealistic, instantaneous horizontal force distribution at that floor, as can be inferred from wall shear force changes from above to below the floor. These force changes are often unrealistically large, implying a need for impractical and unnecessary floor slab strengthening. A better approach to force redistribution through diaphragms is modeling floor slabs at and near outrigger levels as semi-rigid elements, performing sensitivity studies to determine the effects as slab stiffness is varied through realistic stiffness ranges as described in Section 2.5. Typically this will show that force transfers occur more gradually, over several floors, and that force transfers are not very sensitive to diaphragm stiffness assumptions. Gradual force transfers mean the walls aligned with outriggers will have larger forces for several floors.

Modeling Core-Column Interaction
The deflected model in Figure 2.1 in Section 2.2 shows that outrigger columns unconnected to floor slabs will follow straight line paths between outrigger connection points. This imposes no flexure in the columns but is unrealistic. On the other hand, if rigid diaphragms or master/slave nodes are used to link columns to a core, the columns must bend to follow a core at "kinks" or local changes in slope caused by outrigger force couples. Small columns are flexible enough that the moments induced by following core kinks are of no consequence, and the restraint forces needed to provide static equilibrium under those moments are minor. For columns of large cross-section, the forces required to enforce

curves that follow core kinks can become very large – unrealistically so. A suggested approach is to selectively replace rigid diaphragms with semi-rigid (realistic stiffness) diaphragms, starting at the outrigger level and working upward and downward until changes in column shear forces (indicating slab restraint forces) become reasonably small when compared to the rigid-diaphragm model results. Regardless of the diaphragm modeling assumptions, outrigger chords must be designed with sufficient strength to resist at least the full horizontal component of the outrigger diagonal forces. The restraint or bracing force may be taken through chords (where they connect to columns), slab reinforcement or both acting in compatible combination.

Stiffness Ranges
The diaphragm floors of an indirect, virtual outrigger/belt truss system should be analyzed for both gravity loads and lateral loads with a reduced stiffness considering concrete cracking depending on the stress level. Counting on 100% of gross slab properties would be unrealistic. Parametric studies over a range of slab stiffness would be appropriate since different stiffness could apply under different loading conditions. For example, an upper limit of perhaps 50% of gross stiffness would reflect that slabs encircling a core may simultaneously experience compression, shear, and tension in different regions under moderate loads, while a lower limit of the slab reinforcement transformed area would conservatively simulate a case of extensive cracking under extreme loads. For service condition checks such as occupant comfort, acting loads are small and the upper range of effectiveness may be appropriate. For member and connection strength checks, different effectiveness values could be used for different members. For example, core forces are probably worst when slab effectiveness

is low, but slab, belt truss, and participating perimeter column forces are probably worst when slab effectiveness is high. For dynamic properties being provided to a wind consultant, present the range of properties resulting from the range of diaphragm assumptions.

For a direct, conventional outrigger system the outrigger forces should be released from floor diaphragms, even if they are modeled as semi-rigid diaphragm model elements, in a manner that allows all axial force to remain in

Treating floor slabs as rigid diaphragms in analytical structural models is common and computationally efficient, but a rigid diaphragm or master/slave node approach should not be used at or near outrigger floors as it will artificially stiffen the outrigger system, erroneously report zero force in outrigger truss chords, and obscure forces needed to determine force resolution and compatibility with the floor diaphragm.

the outrigger truss members. Once the governing chord forces are determined, sensitivity of building behavior from partial slab participation can be studied to determine if that governs other outrigger members.

Construction

For indirect outrigger or belt truss systems a floor slab of reasonable thickness, in the range of 300 millimeters, is often sufficient to transfer horizontal floor shears from the core to the belt truss, though appropriate thickness must be determined for each floor location of each building. Load transfer approaches from core to slab may include drag reinforcement extending from core walls, and shear friction based on core dowels crossing a roughened or keyed interface, or an integral slab where the construction sequence alternates slab and wall pours. Where a building has steel composite floor framing on typical floors, thick floor slabs at the indirect or virtual outrigger levels can create the floor diaphragm, with drags and dowels engaging the core and extra shear studs transferring floor in-plane shear to the belt wall or belt truss. As an alternative to a thick concrete floor slab, steel horizontal under-floor bracing can provide the necessary diaphragm action if direct connections are provided to the core and the belt truss chords. Both strength and stiffness must be considered. Horizontal bracing located below the floor framing would not interfere with the gravity system, but it may impact headroom and installation of mechanical services and may not align well with belt truss chords. Running horizontal bracing through shop-fabricated web penetrations is another approach. It carries some cost in more complex fabrication and erection, and requires careful study regarding member stability and the interaction of bracing and floor framing from strain compatibility. Such an approach may locate

bracing to better align with chords and minimize interference with mechanical systems running below it.

2.5 Stiffness Reduction

Analytical studies must consider appropriate stiffness regimes depending on the load conditions, especially for concrete construction. When considering concrete core walls and columns, different stiffness reduction factors apply for service-level wind (gross sections), factored wind, and factored seismic cases, as well as further reductions in the presence of cracking, as described in ACI 318. If a nonlinear analysis has modeled explicit changes in member stiffness at different load levels there is no need to apply general stiffness reduction factors as well. However, nonlinear analyses are typically performed only after preliminary member sizing has been performed on the basis of simpler, elastic models. To the extent that geometric nonlinearity (P-Delta effect) is not explicitly considered by the analysis method, lateral stiffness should be reduced to reflect it. Stiffness reduction for realistic diaphragm in-plane analysis is discussed at several points elsewhere in this document. For indirect or virtual outrigger systems, appropriate diaphragm stiffness can be as significant as core and column stiffness. In fact the two are related, as indirect outrigger effectiveness is determined by the relative stiffness of the core compared to the diaphragm, belt, and perimeter column system.

2.6 Differential Column Shortening Effects

In a high-rise building, columns are typically highly strained from gravity loads, and small differences in strain between adjacent columns, or between

columns and the core, will accumulate, resulting in significant differences in axial shortening over a building's height. As outriggers that link columns and the core are displaced by differential movements, the resulting strains can generate very large forces within the outriggers, transferring a portion of gravity loads between columns and core. If no special measures are taken, for some designs gravity transfer forces can be of similar magnitude to the outrigger design forces resisting lateral loads. To avoid having to design for such large forces, or being surprised by potentially damaging forces and displacements in structural and nonstructural elements, consider differential shortening between vertical members throughout the design and construction process.

Initial Proportions

Ideally the gravity system is coordinated with the lateral system so that members of similar materials are used and axial stress levels under gravity are similar for all vertical members. That will minimize differential column shortening. However, in real concrete buildings columns typically have higher axial stresses than core walls and shorten more as a result. The reverse may be true in steel braced core buildings. Outriggers connecting the two types of elements will try to transfer load through the outriggers, for example relieving concrete columns and loading concrete core walls.

Time-Dependent Effects

Time affects outriggers through differential shortening four ways. First, differential shortening at a particular floor any point in time during construction will be affected by the sequence of constructing columns and the core, and the timing of gravity load application as floors are built above. For example, a core may temporarily experience higher gravity strains than perimeter columns if the core is advancing many stories ahead of columns and floor framing. Once framing tops out this condition no longer applies.

Second, foundation dishing can result in differential vertical elevations at core and perimeter columns. Dishing from elastic settlement, such as rock deformation or pile shortening, will increase as building construction proceeds and stabilize as it tops out. If the sub-grade is subject to consolidation settlement, such as clay, dishing may continue to grow for years, at a diminishing rate, as water is gradually squeezed from the sub-grade material. As discussed earlier, dishing may add to or reduce differential vertical shortening effects, and the time range for dishing to develop may be quite different from that for differential shortening to develop. This complicates studies of their potential interaction and influence on the overall frame.

Third, outriggers are affected by differential shortening only after the outrigger trusses or walls are completed. The timing of final connections can establish how much of the total differential shortening has already occurred, and how much differential shortening has yet to occur and affect the outrigger system. This means controlling construction sequence can be an important aspect of outrigger design. Where member forces are significantly affected by construction sequence, the sequence anticipated in the design should be stated within the construction documents. This is further discussed in Sections 2.10, 3.4, & 3.5.

Fourth, buildings with reinforced concrete or composite core walls and columns will experience post-construction strain from creep (continued shortening under constant load) and shrinkage (shortening from concrete drying as it approaches ambient relative humidity)

that typically exceeds strain from elastic shortening. Creep and shrinkage magnitude and timing are affected by the concrete mixture used, ambient relative humidity (which may change from outdoor exposure during construction to conditioned air in service), member volume/surface ratio (higher ratios mean slower rates of creep and shrinkage), and the reinforcing ratio (more embedded steel, whether rolled shapes or reinforcing bars, reduces creep and shrinkage magnitudes). Predicting the magnitude and timing of creep and shrinkage requires accounting for a realistic construction schedule that establishes a history of loading increments, as well as the material and member properties discussed above. Such a staged or sequential construction study can become rather elaborate. At its heart are prediction formulas for elastic, creep, and shrinkage strains, ideally calibrated to laboratory test data on actual concrete mixtures planned for the project. Predicted differential shortening can then be used to investigate possible transfer loads through outriggers, to guide outrigger erection timing and to determine the need for, or effectiveness of, other more elaborate control or mitigation measures such as those discussed in Sections 2.10, 3.4 & 3.5.

Note that time-dependent differential shortening effects are greatest when different materials are used in the core and in the perimeter columns. In a concrete core and steel perimeter design, for example, all post-construction core shortening generates differential shortening. That can be a large number, reaching several centimeters over time.

Temperature Effects
A less common but still significant situation is outriggers connecting members with different thermal exposure, such as perimeter columns exposed to weather. In that case gravity

load shifts can be in either direction depending on the relative temperatures of core and columns, and the load shifts are seasonal. Temperature effects are discussed in more detail later in the text.

Load Combinations
A significant design question is the appropriate treatment of transfer forces in load combinations used for determining required strength. Some model building codes currently show forces from self-strain effects, designated T, in one load combination. Lateral loads from wind or seismic effects are in different load combinations without T. However, this is overly simplified in at least four ways:

▶ A single value for T is not sufficient. Self-strain can have multiple sources, including differential temperatures, concrete creep, concrete shrinkage, and other phenomena. Each source has its own magnitude and timing.

▶ Different load combinations are appropriate for different sources of T. Transfer forces from differential shortening could certainly occur simultaneously with wind or seismic forces, so excluding them from some load combinations is not recommended. This is explicitly stated in recently-issued ASCE 7-10 (ASCE 2010).

▶ Different factors on each of the values of T should apply based on the sources of T and the other forces being considered in each load combination. Where T represents transfer forces based on probable (mean or median predicted) elastic, creep, and shrinkage values it may be appropriate to apply the same load factors as the corresponding gravity forces, such as for dead and live load, since gravity load drives elastic and creep shortening.

Another approach is to establish load factors based on probabilities. For example, when a transfer force acts with gravity alone, consider applying a factor on the probable transfer force to cover a larger (more conservative) predicted shortening value, as would be required to achieve an 85% confidence level (15% probability of exceedance) from historical test data. This can be determined in some creep and shrinkage models. Transfer forces acting in combination with wind or seismic loads could use a load factor of 1.0 on the mean or median prediction values. Load factors and combinations related to self-strain loads T in general and thermal loads in particular are discussed more fully in the following Section 2.7.

▸ Time-based load combinations should be considered. Because gravity load transfer forces vary with time, especially from creep and shrinkage effects, total forces in the core and outrigger columns will vary with time as well. To cover both the immediate and long-term load distribution cases, separate load combinations should be determined both with and without the transfer forces present.

2.7 Thermal Effects Management

Outriggers that link exposed perimeter columns and a temperature-controlled internal core can experience large forces induced by temperature differences. The magnitude of temperature difference should consider realistic heat flow paths, including the ratio of surfaces exposed to the exterior and interior, and the thermal properties of the material. At a minimum the effect should be considered in all load

combinations that include the self-strain load T arising from thermal effects for a realistic range of exterior and interior temperatures. The load factor applied should reflect the probability of occurrence: a larger factor should apply if using seasonal or daily average maximum and minimum temperatures, while a smaller factor could apply if extreme recorded temperatures are used.

ASCE 7-10 (ASCE 2010) is not yet referenced in current codes, but it addresses self-strain load(s) T with general statements. For factored load combinations, it states, "Where applicable, the structural effects of T shall be considered in combination with other loads." Also, "the load factor on T shall not have a value less than 1.0." For load combinations under Allowable Stress Design the wording is identical except for a 0.75 load factor. These statements validate the idea that T should not be limited to selected load combinations, while complicating establishment of appropriate T values. Due to the low probability of simultaneous extreme temperatures and earthquakes or rare winds, a less-than-extreme value for T is recommended so that a load factor of 1.0 (or 0.75 for ASD) is appropriate in combinations with wind (W) or earthquake (E) loads. For combinations without wind or earthquake loads, a higher load factor on T can be applied to cover potential extreme temperatures.

Self-strain loads T can arise from creep and shrinkage, foundation settlement, restrained post-tensioning, and other conditions, in addition to thermal variations. Different self-strain loads can occur at different points during the service life of a building. Temperatures can vary seasonally every year, while creep and shrinkage strains accumulate gradually over years. To address self-strain loads T that arise from different

causes and are subject to different time schedules, separate values and load factors should be established for each condition. For example, thermal effects associated with a load factor of 1.0 (or 0.75) could be based on seasonal daily average temperatures and used in combinations with wind or seismic forces. Higher load factors that reflect thermal effects associated with extreme, 95th percentile high and low temperatures, along with the temperature rise from direct solar heating, could be used in gravity-only combinations. British and Chinese building codes also provide guidance on load factors for T in various load combinations.

2.8 Load Path from Connections

Many building designs use a limited number of outriggers per floor and locate them on just a few floors. This is helpful in minimizing impact on floor usability and the construction schedule. However, outrigger columns often act to resist a large portion of building overturning forces, so each of these outriggers will experience axial forces that are large, varying, and usually reversible. Equally important, those large forces must be transmitted to, and distributed within, the core and the column being joined by the outrigger.

When core, outrigger, and column are all structural steel, the connections will be large but can be conventional. Special details would be needed at special operations, such as bearing pockets for shim stacks or jacks as part of differential shortening management strategies described later in this document.

When forces must transition between different materials, establishing an appropriate load path requires study and creativity; there is no single "correct" approach. Consider, for example, alternatives for the load path from steel outrigger to concrete core wall.

- An embedded plate, flush with the concrete face, can use composite shear connectors ("headed studs") on the plate to resist the vertical component of the force in the outrigger diagonal, while long horizontal bolts developed within the wall can take the horizontal force from a member end plate through nuts on projecting threaded ends. Member end plate and embedded plate sizes should be sufficient to spread compression forces into the concrete. For larger forces this approach may not be practical. Reinforcement within the wall is needed to distribute the bolt tension forces across the wall width (see Figure 2.4). Bolt strain may allow the plate to pull away from the wall, potentially compromising the shear load path and causing concrete degradation. Pre-tensioned high strength rod may minimize or avoid this behavior. Also, appropriate design shear values for headed studs are not obvious; the strength values used for composite beam design are developed only after some

deformation occurs. That may not be appropriate for reversible cyclic outrigger forces.

- Continuous embedded steel members can permit more conventional, direct steel-to-steel connections but have their own drawbacks. Concrete construction is much more complicated when working around heavy steel members, and accuracy of steel placement and

subsequent connection fit up can be affected by the concrete encasement (see Figures 2.5 & 2.6). Design of the embedded steel requires thought: if sized just for strength, to minimize tonnage, will the resulting steel strain be incompatible with surrounding concrete, leading to deterioration? How will forces exit the steel to enter the concrete bond, headed studs, other methods?

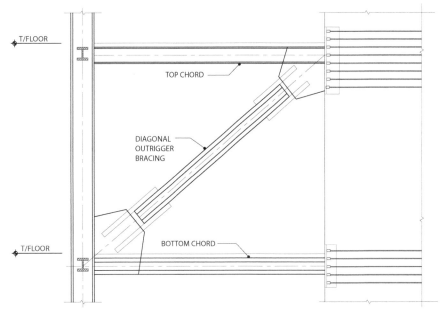

▲ Figure 2.4: Outrigger connections through embedded plates and deformed bar anchors. © Thornton Tomasetti

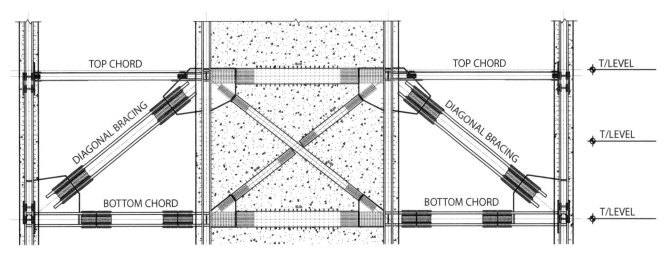

▲ Figure 2.5 Outrigger connections with continuous steel members – full building width view. © Thornton Tomasetti

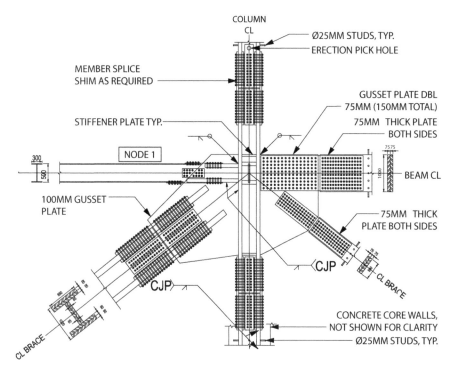

COLUMN
CL

Ø25MM STUDS, TYP.
ERECTION PICK HOLE

MEMBER SPLICE
SHIM AS REQUIRED

GUSSET PLATE DBL
75MM (150MM TOTAL)

75MM THICK PLATE
BOTH SIDES

STIFFENER PLATE TYP.

NODE 1

BEAM CL

100MM GUSSET
PLATE

75MM THICK
PLATE BOTH SIDES

CJP

CL BRACE

CJP

CL BRACE

CONCRETE CORE WALLS,
NOT SHOWN FOR CLARITY
Ø25MM STUDS, TYP.

CL BRACE

▲ Figure 2.6: Outrigger connections with continuous steel members – connection detail (bolting is shown; welding is also possible). © Thornton Tomasetti

▶ Partial height embedded steel members covered with headed studs can use conventional steel-to-steel connections and transfer the force to surrounding concrete along the axis of the steel member (see Figure 2.7). Appropriate design shear values for headed studs, bond, and end plates must be determined. Steel member length depends on the shear transfer values and the forces to be transferred. While partial height members reduce the number of stories that concrete work is affected, headed shear studs on all faces can affect the minimum wall thickness that can fit both steel and reinforcement.

▶ Localized short steel stub members can permit more conventional steel-to-steel connections while limiting impact on concrete construction to the immediate area

(see Figure 2.8). The ability to set and hold the stubs accurately for fit up, and provide a suitable load path are two of the challenges for this approach. A large bearing plate at each end of a stub, sized like a column base plate, can distribute upward and downward forces over the plan cross section of a concrete column or core wall corner or intersection.

▶ Direct bearing details using pockets in concrete may be effective, but may have limited capacity and will impact surrounding reinforcement work.

Concrete to concrete connections may also be complex, depending on outrigger geometry. Transitioning diagonal reinforcing into horizontal and vertical reinforcement, developing bars, lapping bars, and anticipating and resolving different strain values and

patterns for compression and tension in the outrigger member must all be addressed (see Figure 2.9).

2.9 Panel Zone Load Path

When outrigger levels are few and far between, the pattern of shear forces in the core wall is similar to shear in a moment frame column: panel zones, in this case outrigger levels, are locations of larger-than-typical, reverse-direction shears. The core-as-column analogy is not perfect: unlike column webs, core wall panels are typically perforated by lines of doorway openings. Shear stiffness and strength of coupling beams crossing openings may limit "panel zone" capacity. Building designs have addressed this condition several ways.

▶ If openings can be omitted at one or more stories at the outrigger level, it may be practical to design the resulting story-high (or deeper) coupling beam to resist the larger wall shear force. Wall thickness and beam height must be adequate to keep shear stress below code maximum limits, and reinforcing in the beam and adjacent wall panels would be increased to provide the required strength. Note that the coupling beam may be resisting axial load as well as shear and moment if outriggers on opposite sides of the core are loaded in the same direction, as from differential column shortening or thermal changes.

▶ A strut-and-tie model may be applied where opening sizes and locations permit. This requires clear paths with adequate face width and wall thickness for compression struts, bands of continuous reinforcement or embedded tension members for tension ties,

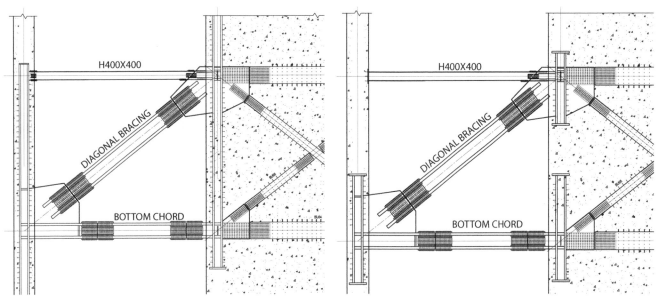

▲ Figure 2.7: Outrigger connections using local embedded steel through headed studs. © Thornton Tomasetti

▲ Figure 2.8: Outrigger connections using steel stubs through studs and end plates. © Thornton Tomasetti

and adequate room at strut/tie intersections for force-transfer nodes. Strut-and-tie can efficiently resolve forces from outriggers on opposite sides of the core acting in opposite directions, as when resisting overturning. Additional tension ties and compression struts are needed for outriggers acting in the same direction. Strut-and-tie modeling can be complicated when outriggers at right angles meet at the same corner of a core. As force magnitudes and directions from those outriggers vary with time, conditions at the nodes will also vary dramatically, requiring special attention. See the Trump Tower discussion in Section 3.3.

▶ If outrigger connections and load paths are already based on embedded steel members, panel zone forces can be resolved through an embedded complete steel truss like that shown in Figures 2.5 & 2.6. This requires favorable geometry, which is also true of the other approaches. With embedded steel,

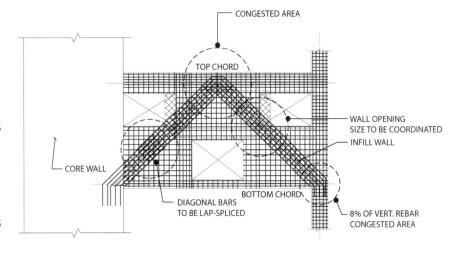

▲ Figure 2.9: Concrete outrigger wall showing bands of reinforcing bars. © Thornton Tomasetti

wall thickness can be minimized – or based on non-panel-zone story shears – and the face width needed for struts and ties can be minimized, allowing for larger wall openings. This approach has its own impacts, including construction schedule delays for erection and more complicated concrete construction procedures to work

around the steel while holding it in position. A "retrofit" construction strategy, casting core walls with blockouts to permit later truss erection and blockout infill, has been used on at least one recent building, Two International Finance Centre, as described in Section 3.5.

2.10 Outrigger System Construction Sequence

Construction of a core-and-outrigger building has two key aspects: mitigation of differential shortening and effect on overall construction schedule. While construction sequence for most buildings may be considered as "means and methods" separate from design, that is not true for outrigger systems due to differential axial shortening effects as mentioned in Section 2.6. No matter which structural systems and materials have been selected by the designer, time dependent shortening effects cannot be neglected or eliminated. They can, however, be reduced through construction sequence.

One aspect of time dependent shortening is the gradual application of load during the construction process. The portion of time dependent shortening that occurs before an outrigger is erected at a particular floor will not generate gravity load transfer effects. This effect can be determined through a staged analysis using a series of partial-height computer models. Delaying final outrigger connections can provide a further reduction in load transfer effects. Generally speaking, by the time a structure has been topped out, up to 95% of elastic shortening has occurred. Remaining elastic shortening comes from the weight of finishes, building services, and live load.

The other, more challenging aspect of time dependent shortening is the significant axial shortening that will continue to occur at concrete core walls and concrete columns from creep and shrinkage. This means that delaying final outrigger connections linking core and columns until after topping out can reduce, but not eliminate, additional forces in the outrigger system since differential post-top-out axial shortening of core and columns will occur.

These forces must be considered in the design. Minimizing these additional forces would preserve more of the outrigger capacity to resist lateral forces. This requires a system that allows later adjustment of the outriggers, connections. Several such systems have been proposed and some have been implemented in recent tall buildings.

Shim Plate Correction Method

Where core-alone (or core plus frame) strength and stiffness is not sufficient for construction-period lateral loads, steel truss outrigger connections to perimeter framing can incorporate shim packs. Outriggers act by compressing the shim packs. Connections are monitored during and after the construction period so shims can be removed or added as required to compensate for gradual differential movements between outrigger ends and columns (see Figure 2.10). As a gap decreases on one side of the outrigger end and increases on the other side, shims are relocated to maintain the gaps within a specified range. If a gap closes before shims can be repositioned, an oil jack can move the outrigger end enough to release the shims for repositioning. The approach requires more labor and monitoring than conventional details, as well as the use of monitoring devices (Chung et al. 2008). Shim plate correction offers the potential for greatly reducing unwanted load transfer forces, but could result in unanticipated forces if monitoring or maintenance of the specified gaps fails to occur. This approach was used on Cheung Kong Centre discussed in more detail in Section 3.4.

Oil Jack Outrigger Joint System

Another concept to minimize differential shortening forces during and after construction, while avoiding the need for continual monitoring and periodically shifting shims, uses pairs of oil jacks and a connecting pipe with an orifice (see Figure 2.11). Top and bottom cylinder jacks would be installed and filled with oil to extend rams for tight contact with outrigger ends. Closing the supply valve would establish the total volume of oil in two jack cylinders and connecting pipes. Resistance to flow through an orifice is proportional to velocity raised to a power. Movement from differential shortening through months or years would allow oil to flow from one cylinder to the other, generating very little resistance. Movements taking seconds during wind storms or earthquakes would resist oil flow, making the outriggers fully effective against overturning moments and deformations (Chung et al. 2008). In this system, reliance on jacks, pipes, and orifices can be supported only with a program of inspection and maintenance, and a degree of redundancy that includes sensitivity studies for safe and acceptable building behavior under design loads with the loss of one or more jack sets.

Cross Connected Jack System

Another proposed system to allow for differential shortening during construction while maintaining outrigger effectiveness against overturning uses cross-connected oil-filled flat jacks with the top face of one outrigger hydraulically tied to the bottom face of the outrigger on the opposite side of the building (see Figure 2.12). This approach avoids the need for oil flow control through an orifice. Under uniform vertical movement total hydraulic volume is constant as one jack closes and the other opens. Overturning tries to close both jacks, increasing pressure in the system which resists the moment. To avoid relying on the hydraulic system permanently, the jacks would be grouted late in the construction process (Kwok & Vesey 1997). The system would be effective for outriggers aligned with wind load direction, but would not provide the benefit of smaller, but

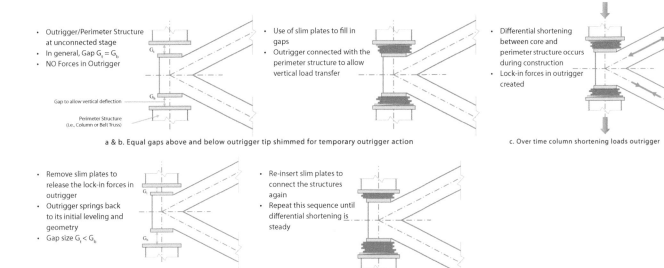

- Outrigger/Perimeter Structure at unconnected stage
- In general, Gap $G_t = G_b$
- NO Forces in Outrigger

Gap to allow vertical deflection

Perimeter Structure (i.e., Column or Belt Truss)

- Use of slim plates to fill in gaps
- Outrigger connected with the perimeter structure to allow vertical load transfer

- Differential shortening between core and perimeter structure occurs during construction
- Lock-in forces in outrigger created

a & b. Equal gaps above and below outrigger tip shimmed for temporary outrigger action

c. Over time column shortening loads outrigger

- Remove slim plates to release the lock-in forces in outrigger
- Outrigger springs back to its initial leveling and geometry
- Gap size $G_t < G_b$

- Re-insert slim plates to connect the structures again
- Repeat this sequence until differential shortening is steady

d. Jack to relieve loads, remove shims

e. Shim again in new position to restore outrigger action

▲ Figure 2.10: Shim plate correction method to release differential shortening forces. © Arup

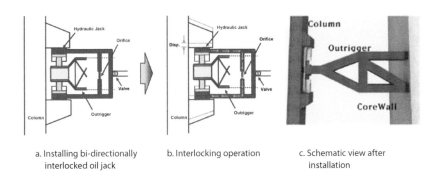

a. Installing bi-directionally interlocked oil jack

b. Interlocking operation

c. Schematic view after installation

▲ Figure 2.11: Oil jack outrigger joint system. (Source: Chung et al. 2008)

still potentially useful, reactions from outriggers perpendicular to the wind load direction; because they would be moving in the same direction on opposite building faces, the perpendicular outriggers would act as released, not as locked.

2.11 Code Interpretations for Seismic Load Resisting Systems

Current seismic design provisions in building codes, such as the International Building Code (IBC) and Eurocode 8, were not developed for application to tall buildings since they comprise a small portion of overall building construction. Prescriptive seismic design provisions in these building codes do not sufficiently address many facets of seismic design of tall buildings, such as the outrigger systems frequently used for lateral load resistance of tall buildings; they are not currently included as an option under the Basic Seismic-Force Resisting System table in the IBC. In addition, many building codes have height limitations on many practical and popular seismic force resisting systems, which block their use in tall structures if following prescriptive provisions.

The CTBUH has prepared guidelines addressing the issue of seismic design of tall buildings (Willford et al. 2008). It presents the most appropriate approach as being performance based design (PBD) rather than prescriptive design. This makes sense for several reasons. The United States' prescriptive design is based on nonlinear response through system ductility, as response

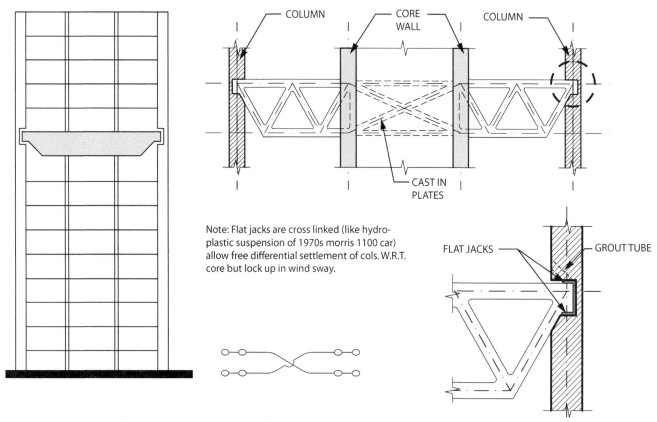

Note: Flat jacks are cross linked (like hydro-plastic suspension of 1970s morris 1100 car) allow free differential settlement of cols. W.R.T. core but lock up in wind sway.

▲ Figure 2.12: Cross-connected jacks during construction. (Source: Kwok & Vesey 1997)

modification factor **R** is used to reduce elastic response to a design level. However, the lateral load resisting system of a tall building can have different structural elements with very different ductility capacities and demands: coupling beams between core walls may experience high demand from story shears driven by participation of higher modes and require high ductility, while outriggers may experience proportionately lesser demand from overturning moment that is smaller than indicated by story shear and be designed for elastic or limited-ductility behavior. For tall buildings, minimum base shear is typically greater than shear determined by response spectrum and building period. As a result, the effective **R** value in prescriptive design is reduced

from the value specified in the building code for that framing system to a smaller value dependent on building period and other factors (Moehle 2007). Also, prescriptive procedures may underestimate shear demand and may not provide the required flexural ductility at the core base (Willford & Smith 2008).

PBD, as permitted in the code as an alternative to prescriptive design, offers clear benefits for achieving better tall building designs. It requires clearly defined performance objectives, procedures for selecting and scaling earthquake ground motions for design, nonlinear modeling methods that produce reliable estimates, acceptance criteria for calculated demands and a framework for the design and review of

alternative-design buildings (Moehle 2007). For outrigger systems used for the lateral load resistance of tall structures, performance-based analysis during design is frequently requested by reviewers.

Criteria different from prescriptive codes can be reflected in PBD. Outrigger members can be designed to remain elastic under the Design Basis Earthquake or the Maximum Considered Earthquake, or can be "fused" to limit the member forces and absorb seismic energy. Which strategies to apply should be discussed between designers, owners, reviewers, and governing jurisdictions at the start of the design process.

2.12 Soft-Story and Weak-Story Seismic Requirements

Prescriptive code seismic requirements limit the permissible variation in stiffness or strength from story to story. In particular, codes discourage having a stiffer or stronger story above a softer or weaker story. This would appear to prohibit outrigger systems, since the story below an outrigger is usually significantly less stiff and weaker in shear than the outrigger story. However, such an objection would be short-sighted: the code requirement is intended to guard against a uniformly stiff or strong building having a soft or weak-story where deformations would be concentrated, as may occur at a lobby or other non-typical level. Outriggers create the opposite situation. A building with multiple stories of similar stiffness and strength is additionally strengthened and stiffened at the few outrigger floors. Having many similar floors provides ample opportunities for well-distributed ductile behavior between outriggers, while the outriggers provide positive global effects. Performance-based analysis can demonstrate that behavior is acceptable under this system.

2.13 Strong Column Seismic Requirement and Capacity Based Design

Outriggers must be stiff and strong when linking a core to perimeter columns to be effective in restraining building flexural deformation and base overturning. However, the effects of stiffness and strength must also be considered for building behavior during large earthquakes. They could generate short-term forces large enough to damage the outrigger, the columns, the core, or connections. A typical seismic design requirement is to avoid loss of gravity system capacity, making failure of columns and cores of particular importance. For example AISC Seismic Provisions (AISC 2002) require that columns have axial compressive and tensile strength, *in the absence of any applied moment,* [emphasis added] to resist the least of load combination forces using amplified seismic load, force delivered by other members acting at capacity, or the foundation resisting uplift. The intent of this requirement can be met three ways: by demonstrating gravity support capacity is maintained in the event of overload, by keeping capacity greater than demand, or by using capacity based design to control demand on some elements by limiting the capacity of other elements.

Maintaining capacity in the event of overload – in effect, "ductile" axial behavior – may be impractical. Concrete column axial ductility could require transverse confinement reinforcing of hoops and crossties comparable to those in special shear wall boundary zones and special moment frame beams and columns, the amount of transverse reinforcement needed to confine high-strength mega columns would be daunting. For steel columns, keeping slenderness so low that squashing rather than buckling controlled would require very thick plates. With such heavy members it is likely that other elements would yield first. So amplified load checks and capacity based approaches are more suitable for practical design.

The intent of the amplified seismic load requirement may be met in some designs by performance based analysis. For lateral systems sized for other criteria, such as stiffness and strength under extreme wind loads, nonlinear time history studies may be able to demonstrate that demand under load combinations including seismic effects never exceeds capacity at outriggers and the columns to which they connect, even for the maximum considered earthquake event. This situation is more likely to occur at outriggers and mega columns near mid-height: gravity load will comprise a large portion of the column axial demand, column net tension is less likely to occur there than at outriggers high in the building, and outriggers may be designed for large forces that include gravity load transfers between core and columns.

Capacity based design can avoid the need for highly ductile column axial performance by limiting applied forces from seismic events to a maximum value in combination with well established factored gravity forces from dead and live loads. The capacity-based approach to avoiding column failure relies on having non-column members yield or buckle first. Establishing outrigger members small enough to serve as "fuses" may be achievable by optimization. Where the lateral load resisting system is being sized for stiffness, as may occur where wind criteria are governing the design, the

Outrigger members can be designed to remain elastic under the Design Basis Earthquake or the Maximum Considered Earthquake, or can be "fused" to limit the member forces and absorb seismic energy.

▲ Figure 2.13: L. A. Live Tower. © Nabih Youssef Associates

serves to transfer gravity load as well as outrigger load, or to perform as an indirect or virtual outrigger in parallel with the direct outriggers.

▶ Some designs have included connection details capable of slipping at defined values. Such an approach must consider service-level and strength-level forces from wind loads compared to seismic loads, slip distances needed to provide sufficient seismic force relief, variability of slip characteristics, permanent deformations after a quake, and behavior if the bolts bottom out in their slots.

▶ Linked pairs of hydraulic jacks between outriggers and columns could permit slow outrigger movement to minimize gravity load transfers between core and columns, as discussed for the Oil Jack Outrigger Joint System earlier in this document, while resisting rapid outrigger movements from wind and seismic deformations. The linked jacks could support capacity based design through pressure relief valves that allow fluid to bypass the usual resistance orifices once force reaches a preset value. Such an approach would need other backup systems, such as yielding members, in case the valves malfunctioned.

requirement can be met by a variety of combinations of column stiffness and outrigger stiffness. For example, making columns larger to resist capacity-based outrigger forces will add to system stiffness. That may permit downsizing the outrigger members themselves while meeting required system stiffness. The smaller outrigger members would limit the maximum force columns could experience. Since core-and-outrigger systems are indeterminate, changing the outrigger and mega column stiffness will also change the forces they attract. Several design cycles may be required to simultaneously achieve the required stiffness and a hierarchy of strength.

Where redistribution of member stiffness is not sufficient to result in conventional non-column members limiting the forces generated, other strategies must be considered. Force-limiting concepts for insertion between outriggers and columns have been proposed.

▶ Where the load path from outrigger to column is indirect, through a belt truss, it may be practical to use belt truss member capacities to limit the force delivered. However this situation is not that common, and before deciding to permit post-yield behavior one must consider whether the belt truss

The most direct approach to capacity based design is to make the outriggers themselves the "fuses," with outrigger diagonals comprised of Buckling Restrained Braces (BRBs). BRB members have inner steel plates of controlled dimensions and material properties, coated by bond-breaker material and surrounded by a concrete-filled steel jacket. This way BRB member limit states are based on ductile yielding of the inner plates in tension and compression, rather than non-ductile buckling of a conventional member. Each structure must be individually evaluated for BRB practicality. This approach may be particularly effective where wind stiffness is not governing the outrigger design, so that the inner plates can be sized for strength alone. The BRB approach offers several advantages. First, the capacity of each member in both tension and compression is designed, fabricated, and tested within tight elastic and plastic behavior limits so adjacent members are protected against unanticipated overload forces. Second, the BRB avoids buckling that can cause serious strength degradation in relatively few load cycles so the brace remains functional during extreme events. Third, controlled tension and compression yielding can absorb considerable energy, improving overall building performance during a quake. Fourth, BRBs are detailed for easier replacement than conventional members to restore system strength and alignment after a major event.

The 54-story L.A. Live tower built in 2008 (see Figure 2.13) has a steel structural frame with steel plate shear walls within the core (see Figure 2.14). To improve lateral stiffness, perimeter columns are engaged by outriggers at mid-height and at top levels. To avoid overloading those columns in a major earthquake, outrigger diagonals are BRBs (see Figure 2.15). In this design, braces were sized to remain elastic under factored wind loads, which are comparable to forces under the design basis earthquake (DBE). Yielding to limit column forces would occur only under the 50% larger maximum considered earthquake (MCE) (Youssef et al. 2010).

The Russell Investments Center, originally WaMu Center/Seattle Art Museum Expansion completed in 2006 has a tower with a 186-meter-tall concrete core only 9.5 meters wide (see

▲ Figure 2.14: L.A. Live Tower – steel plate shear walls. © Nabih Youssef Associates

▲ Figure 2.15: L.A. Live Tower – Buckling Restrained Braces (BRBs). © Nabih Youssef Associates

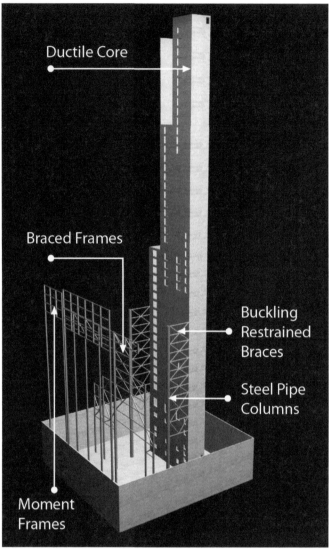

Ductile Core

Braced Frames

Buckling Restrained Braces

Steel Pipe Columns

Moment Frames

▲ Figure 2.16: Russell Investments Center, Seattle. © Benjamin Benschneider (Courtesy NBBJ)

▲ Figure 2.17: Russell Investments Center – BRBs layout. © MKA

Figure 2.16). To improve its stiffness, the engineers used 44 BRBs over 13 stories to engage separate concrete-filled steel pipe columns (see Figure 2.17). The BRB limit maximum forces acting on the columns, and overall building behavior was verified under Performance Based Design through nonlinear time history analyses (Loesch 2007). One Rincon Hill South Tower in San Francisco also uses BRB outriggers to advantage, as discussed in Section 3.4.

2.14 Strong Column Weak Beam Concept in Outrigger Systems

As with soft-story and weak-story provisions, a "strong column, weak beam" building code seismic provision can be misapplied to tall building outrigger systems. The strong column, weak beam provision, called the column-beam ratio in AISC Seismic Provisions and minimum flexural strength of columns in ACI 318 Seismic Provisions, specifically refers to special

moment frames, checking that lateral loads will cause yielding in beams rather than in columns. It is intended to avoid hinge formation in multiple columns at the same level, which could cause story collapse. By requiring column flexural strength to be greater than beam strength at each joint, the provision aims for columns to act as continuous spines for the full height of the structural frame. This way the moment frame beams at many floors must yield and form hinges, absorbing a large amount of seismic energy before collapse can occur.

Applying the strong column, weak beam provision to a core-and-outrigger system building is inappropriate, problematic, and unnecessary. It is inappropriate because outriggers are not moment frames. It is problematic because any realistic outrigger truss or outrigger wall viewed as a "beam" connected to the outrigger column at top and bottom chord levels will not yield in flexure (chords yielding or buckling) before the column does. Outriggers with enough stiffness and strength to be effective will provide a force couple greater than column flexural capacity. Aiming for outrigger chord yielding may also be counterproductive where the chords are bracing the column. There are some strategies to minimize moments that can act on columns, including framing outriggers to belt trusses that can flex transversely and detailing outriggers to load columns at single points, as shown in Figures 2.10 and 2.11 in Section 2.10. However the strong column, weak beam provision does not appear necessary at perimeter columns. In a core-and-outrigger system where the core itself provides the majority of inter-story stiffness, it is evident that story collapse should not occur even if columns develop flexural hinges. By that logic, a strong column, weak beam criterion should apply only for viewing the core as a "column" and the outriggers as "beams" because the central core walls or core braced bays will provide the strong spine desired for favorable seismic performance. Even if perimeter columns hinge at outrigger top and bottom chord levels, the story cannot collapse as long as the core is standing.

While the strong column weak beam criterion is inappropriate at perimeter columns, it does make sense when looking at the core as the "column" and

the outrigger trusses as the "beams," checking that the outriggers do not develop forces large enough to cause core failure is both rational and practical. Therefore, the performance based seismic design approach for outrigger system buildings is highly recommended, looking at responses to realistic seismic events. A PBD approach can demonstrate that capacity-limiting measures such as BRB diagonals work. Alternatively, the need to design an outrigger as a "weak beam" can become moot if members sized for strength and stiffness are shown to remain elastic in a nonlinear time history response analyses and the core can resist the resulting forces.

Even if strong column, weak beam criteria for column hinging are not applied, column axial strength provisions against failure in compression must still be satisfied as described above. Also, the influence of column flexural strain on framing design should be considered, including the possibility of cumulative damage from compressive strain through multiple cycles of axial and flexural loading.

Another significant difference between moment frames and core-and-outrigger systems is the ability to manage deformations through design. For example, where a stiff outrigger system can apply local shear and moment forces large enough to cause a significant local change in inter-story drift, in effect the core is locally "kinked" as seen in Figure 2.1 in Section 2.2. If framing members tying a mega column to the core are very stiff, they can force the column to "kink" as well, generating large column moments and large horizontal bracing forces to generate those moments. In effect, the outrigger framing acts like a moment frame joint, enforcing compatible rotations. But if members connecting to the mega column have less axial stiffness, their

axial strain can significantly reduce the forces and kink effect acting on the column.

2.15 Capacity Based Connection Design

A general seismic design principle is to have connections stronger than members. The intent is to maximize ductile behavior by distributing post-yield strains along as much of the member length as possible, rather than having yielding, and potential fracture, concentrated within the connections. For massive outrigger members sized to satisfy stiffness requirements, it may not be practical to provide connections stronger than the maximum capacity of the member. In such cases the results of nonlinear time history analyses as part of a PBD approach can be used to determine realistic connection demand. The connections can be designed to resist the demand from unreduced seismic conditions while staying elastic, or with limited "hot spots" of yielding.

In a core-and-outrigger system where the core itself provides the majority of inter-story stiffness, it is evident that story collapse should not occur even if columns develop flexural hinges.

3.0 System Organization and Examples

3.0 System Organization and Examples

3.1 System Development

As core-and-outrigger systems were developed in the 1980s and 1990s, it became clear that core stiffness was critical to successful outrigger systems. While cores can be steel braced frames or concrete shear walls, concrete provides stiffness economically while providing fire-rated separations. In contrast, steel core columns sized for stiffness can grow large enough to adversely affect space planning where they protrude into corridors and elevator hoistways. Large central cores encompassing elevator shafts and stair wells, combined with the development of higher strength concretes and high-rise forming and pumping technologies, have led to concrete as the dominant choice for core structures in very tall towers employing outriggers today. Another widely-used approach is composite construction, with continuous steel columns embedded within concrete columns and sometimes in core walls as well. Composite construction will typically be more expensive than conventional reinforced concrete construction, but offers benefits that include smaller plan dimensions of columns and walls, reduced creep and shrinkage, direct, reliable steel-to-steel load paths at connections, and the means to distribute forces into concrete encasement gradually rather than all at once at the connection.

For supertall towers using outrigger systems without a complete perimeter moment frame, a large core size is critical to provide great building torsional stiffness since the exterior frame contributes relatively little. Wind tunnel testing and monitoring of actual occupied tall buildings has confirmed that torsional motions have potential for being the most perceived by building occupants, so torsional stiffness for motion control can be important.

Horizontal framing is also a consideration in outrigger systems, as outrigger truss chords that are deeper and heavier than typical floor framing can affect headroom below and may lead to non-typical story heights to compensate for such conditions.

Core-and-outrigger systems can generally be categorized based on their structural material. Examples of various system assemblies in the following section highlight the ways the core-and-outrigger system has been adapted to a wide variety of building types and architectural design concepts, including some of the tallest towers in the world, both constructed and proposed.

As core-and-outrigger systems were developed in the 1980s and 1990s, it became clear that core stiffness was critical to successful outrigger systems.

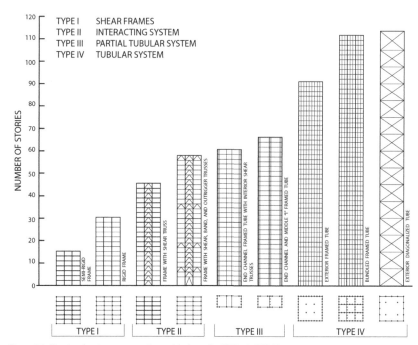

▲ Figure 3.1: Structural systems comparison table from the 1970s © CTBUH

3.2 All-Steel Core-and-Outrigger Systems

U.S. Bank Center (formerly First Wisconsin Center)
Milwaukee, USA

One of the first examples of the system as configured in steel is the 42-story U.S. Bank Center in Milwaukee completed in 1973 (see Figure 3.2). Engineers at the time termed the system a "partial tube." Indeed, the system charts developed at the time indicated the core-and-outrigger system as being applicable only to mid-rise buildings (see Figure 3.3). They considered that outriggers extended the useful range of core-alone systems only marginally. This underestimated their effectiveness for ever taller towers.

The system was selected by the engineers and architects to "create a light open-frame type structure on

▲ Figure 3.2: U.S. Bank Center, Wisconsin. © Marshall Gerometta/CTBUH

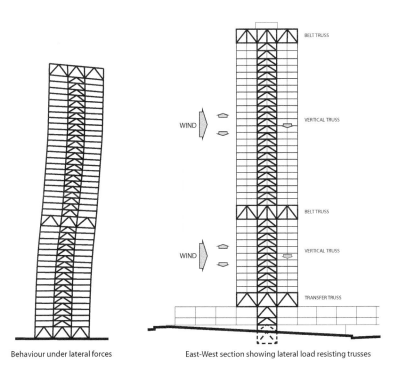

Behaviour under lateral forces

East-West section showing lateral load resisting trusses

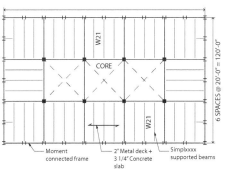

Typical floor framing plan

▲ Figure 3.3: U.S. Bank Center – structural diagrams. (Source: Beedle & Iyengar 1982)

▲ Figure 3.4: New York Times Tower, New York. © Marshall Gerometta/CTBUH

the exterior with columns six meters apart along the perimeter. The frame is continuous with the belt trusses which are expressed architecturally on the exterior." The structural organization was consistent with some key system features still used today: stiff two-story deep outrigger trusses placed at the mechanical levels, linked with belt trusses in order to engage all of the columns in the resistance to lateral loads. The engineers reported a 30% increase in overall lateral stiffness through the utilization of the outrigger and belt trusses.

New York Times Tower
New York, USA

The New York Times Tower is a 52-story addition to the Manhattan skyline completed in 2007 (see Figure 3.4). The large 20 by 27 meters braced steel core is linked to the perimeter through outrigger trusses at the 28th and 51st floor mechanical levels (see Figure 3.5). Columns are typically 9.14 meters on center along the perimeter and some columns are exposed to weather. An important feature of the outrigger system is the potential for redistribution of gravity load between the core and the perimeter frame, making construction sequence important for accurate load sharing predictions through sequential or staged computer analysis. A unique feature of this design was the use of "thermal outriggers" to redistribute thermal strains, minimizing differential strain between columns by reducing the strain of exposed perimeter steel columns while engaging and straining interior columns. This adds to outrigger design forces but reduces floor slopes between the columns to acceptable levels under temperature extremes (Scarangello et al. 2008; Callow et al. 2009; SINY 2006).

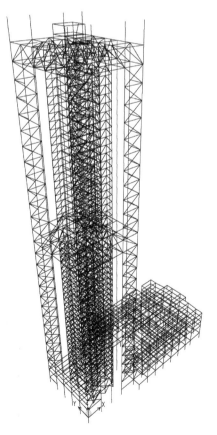

▲ Figure 3.5: New York Times Tower– lateral system. © Thornton Tomasetti

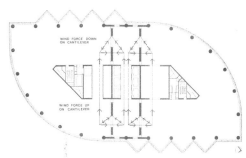

▲ Figure 3.7: Waterfront Place – outrigger plan. © Bornhorst & Ward

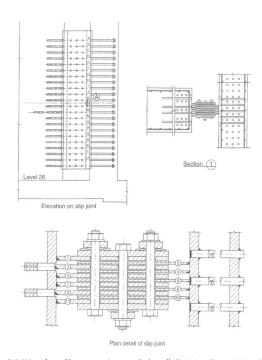

▲ Figure 3.8: Waterfront Place – outrigger to belt wall slip joint. (Source: Kowalczyk 1995)

▲ Figure 3.6: Waterfront Place, Brisbane. © Brett Taylor

3.3 All-Concrete Core-and-Outrigger Systems

Waterfront Place
Brisbane, Australia

An early innovative example of structural engineers addressing the issue of gravity load transfer through stiff outrigger elements can be found in the Waterfront Place project in Brisbane (see Figure 3.6); completed in 1990. The 40-story tower is framed entirely in reinforced concrete, with the core walls linked to the perimeter columns through two-story-tall outrigger walls between Levels 26 and 28 (see Figure 3.7). As the perimeter column lines do not line up with the core walls, outrigger walls are connected through belt walls on the perimeter, which in turn connect to exterior columns.

Two noteworthy features of the design represent pioneering approaches to the outrigger design of reinforced concrete towers. First, the transfer of gravity load between the outrigger walls and the perimeter belt walls was mitigated, but not completely eliminated, through a sliding friction joint at the intersections of these walls. The clamping force in the joint allowed for adjustment to slip at the design load transfer (see Figure 3.8). The joint was then locked down for the remaining life of the structure, differential shortening effects from subsequent live load and superimposed dead load still act on the outrigger. Second, the large openings required through the outrigger walls required the use of extensive strut-and-tie modeling of these elements. Such modeling has

▲ Figure 3.9: Two Prudential Plaza, Chicago. © Marshall Gerometta/CTBUH

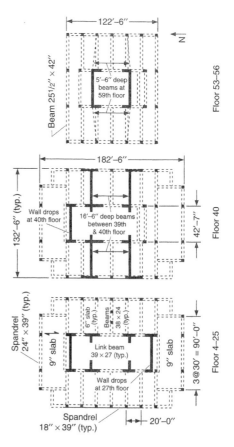

▲ Figure 3.10: Two Prudential Plaza plans showing outrigger walls. (Source: Kowalczyk 1995)

become commonplace today in the design of large, deep heavily reinforced elements like these walls.

Two Prudential Plaza
Chicago, USA

An alternate solution to the concrete outrigger construction was employed for the Two Prudential Plaza tower in Chicago, which was also completed in 1990 (see Figure 3.9). The 303-meter-tall tower has two sets of reinforced concrete outrigger walls at Levels 40 and 59 (see Figure 3.10). These walls are five meters deep at Level 40 and 1.7 meters deep at Level 59. 12,000 psi (85 MPa) concrete is introduced in this design. To reduce transfer of gravity load between the core and the perimeter through the stiff outrigger walls, a short section of outrigger walls

adjacent to the perimeter columns was left temporarily un-concreted for a period during construction. After a suitable time had elapsed and before the construction team de-mobilized, the blocked-off section of the outrigger wall connection was concreted. At the time of completion, Two Prudential Plaza represented one of the tallest concrete structures in the United States and certainly one of the tallest core-and-outrigger designs in the world.

Millennium Tower
San Francisco, USA

The particular challenge of providing sufficient strength, stiffness, and ductility in a design for high seismic demand was realized in the design for the Millennium Tower in San Francisco, completed in 2008 (see Figure 3.11).

This 58-story residential tower was the tallest all reinforced-concrete tower in the western United States at the time of completion. In the short direction of the building, the central core walls are connected to the perimeter at three locations along the tower shaft. Two lines of outriggers connect the core to four "super-columns" located in line with the outrigger walls (see Figure 3.12). Each connection between the core and the super-columns involves five-story punctured wall elements which allow for passage of residents through the perforations in the outriggers. The outriggers are comprised of a combination of heavily reinforced wall elements and diagonally reinforced coupling beams (see Figures 3.13). The capacity-based design approach of ACI 318, Section 21 was used to design

▲ Figure 3.11: Millennium Tower, San Francisco. © Hydrogen Iodide

▲ Figure 3.12: Millennium Tower core-and-outrigger. © DeSimone Consulting Engineers

Each connection between the core and the super-columns involves five-story punctured wall elements which allow for passage of residents through the perforations in the outriggers.

▲ Figure 3.13: Outrigger wall construction. © DeSimone Consulting Engineers

▲ Figure 3.14: Trump International Hotel & Tower, Chicago. © Marshall Gerometta/CTBUH

outriggers, outrigger coupling beams, outrigger connections to core walls, and super columns in the region of the outrigger connection (Derrick & Rodrigues 2008).

Trump International Hotel & Tower Chicago, USA

The Trump International Hotel and Tower completed in 2009 is currently the tallest all-concrete tower in North America, and the tallest built since the Willis Tower (formerly Sears Tower) in the mid-1970s. The 92-story tower is very slender in the short direction with an overall height to least width aspect ratio of approximately 8 to 1 (see Figure 3.14). Massive outrigger wall beams 1.7 meters wide and 5.3 meters deep also serve as column transfer elements for a series of architectural setbacks along the tower height. The outrigger wall-beams are highly reinforced with 75,000 psi (520 MPa) reinforcing bars, dead-end anchorages, mechanical couplers instead of lap splices, and a high performance concrete mix design involving 16,000 psi (110 MPa) self-consolidating concrete. Strut-and-tie modeling was employed for the outrigger elements, and some particularly highly stressed elements required the introduction of 70,000 psi (520 MPa) steel plate reinforcing (Baker et al. 2009; Baker et al. 2006).

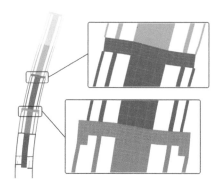

▲ Figure 3.15: Core-and-outrigger diagram (Source: Baker et al. 2006)

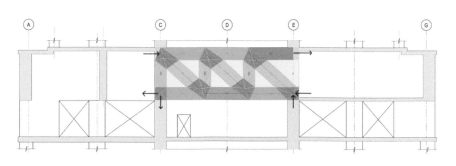

▲ Figure 3.16: Strut-and-tie outrigger link beam diagram (Source: Baker et al. 2006)

▲ Figure 3.17: Plaza 66, Shanghai showing mechanical/outrigger levels as bands. © H.G. Esch/ Kohn Pedersen Fox

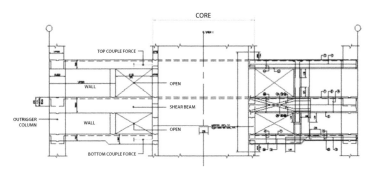

▲ Figure 3.18: Plaza 66 – Outrigger detail. (Tomasetti et al. 2001) © Thornton Tomasetti

▲ Figure 3.19: Plaza 66 – Outrigger Interior view. © Thornton Tomasetti

The Trump Tower design did not include any construction-related postponement or special connections to directly mitigate the issue of gravity load transfer through the outrigger elements. From special analyses of time-dependent shortening effects, the wall designs incorporated additional forces determined through full three-dimensional sequential analysis. Outrigger and belt walls constructed as the tower progresses vertically restrain a portion of the vertical differential shortening between the center of the tower and the perimeter (see Figures 3.15 & 3.16). The high mass and damping of concrete framing helped tower performance in limiting motion perception by building occupants without supplementary damping.

Plaza 66
Shanghai, China

The 66-story, 288-meter Plaza 66 tower in Shanghai was the tallest concrete building in Shanghai at the time of completion in 2001. The versatility of the core-and-outrigger system, even when applied to an area of moderate to high seismicity, was again proven by the design. Similar to the Brisbane and San Francisco designs outlined previously, perforated concrete outrigger elements occur at three mechanical zones (see Figure 3.17). The top and bottom floors of the two-story outrigger frames provide the tension and compression force couple generated by the outrigger effect, while the middle level transfers a good deal of the vertical shear (see Figures 3.18 & 3.19). Six lines of concrete outriggers are employed in the system (Tomasetti et al. 2001).

> The top and bottom floors of the two-story outrigger frames provide the tension and compression force couple generated by the outrigger effect, while the middle level transfers a good deal of the vertical shear.

▲ Figure 3.20 Dearborn Center, Chicago. © SOM

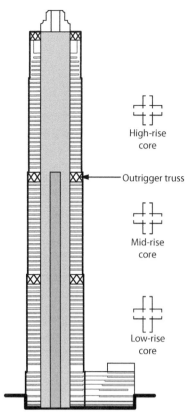

▲ Figure 3.21 Dearborn Center – design elevation. © SOM

3.4 Mixed Steel-Concrete Core-and-Outrigger Systems

Dearborn Center (Proposal) Chicago, USA

Starting in the early 1980s, there was an exploration of options for using simple, stiff reinforced concrete cores with steel long-span floor beams and perimeter columns. An early proposal for such a system was the 85-story Dearborn Center in Chicago (see Figure 3.20), where the architectural team wished to create significant shaping involving horizontal and vertical offsets in the façade. The engineers proposed a cruciform shaped core (see Figures 3.21) organized around the elevator shafts with two steel outrigger lines in each direction at three locations along the building height. At this early stage in the development of the core-and-outrigger system, the engineers for this project recognized the importance of a strong and stiff core as an essential ingredient in the overall efficiency of the system in resisting lateral loads.

One Rincon Hill South Tower San Francisco, USA

For the 180-meter, 64-story One Rincon Hill South Tower in San Francisco completed in 2008 (see Figure 3.22), the design team proposed an alternative to the concrete outrigger walls with punched opening used in the previously described Millennium Tower. The two towers are similar, with two outrigger lines and four large perimeter columns opposite the core, but the outriggers for One Rincon Hill are steel K-braces extending over four stories which allow for significant openings through the outrigger trusses (see Figure 3.23). The braces themselves are buckling restrained braces (BRBs) selected to control the maximum seismic demand on tower columns by yielding in an extreme event. Yielding of the braces is anticipated only for major seismic events, so the BRBs do

▲ Figure 3.22: One Rincon Hill South Tower, San Francisco. © MKA

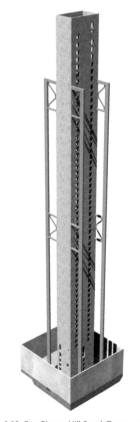

▲ Figure 3.23: One Rincon Hill South Tower – core-and-outrigger system with buckling restrained brace (BRB) diagonals in red. © MKA

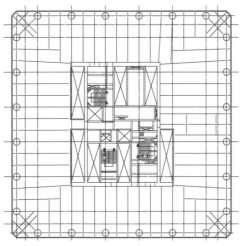

▲ Figure 3.25: Cheung Kong Centre – typical floor plan. © Arup

▲ Figure 3.24: Cheung Kong Centre, Hong Kong. © Arup

▲ Figure 3.26: Cheung Kong Centre – exterior view of the outrigger end at belt truss mid-bay. © Arup

not contribute to damping under wind events. Supplementary damping from roof-level water tanks is provided for occupant comfort (Nolte 2006). Because outrigger systems are not recognized by a prescriptive approach to seismic design, and a prescriptive dual system was not provided, the tower was designed and approved by the local building authorities using performance based design (PBD).

Cheung Kong Centre
Hong Kong, China
This 283-meter-tall high-rise building completed in 1999 consists of a core

with reinforced concrete walls and a perimeter tube with concrete filled tube columns at six meters center to center spacing on all four elevations (see Figures 3.24 & 3.25). Steel outriggers link the core and perimeter. The rectangular core has a maximum height/width ratio of 15, making outriggers important to meet building lateral strength and stiffness criteria during and after construction. Outriggers along wall planes do not align with perimeter columns, framing instead to dedicated outrigger connection points on a belt truss system that minimizes the shear lag effect and evenly distributes

outriggers vertical forces to perimeter columns.

If not released, differential shortening forces could have been as large as design wind forces. To release differential shortening forces but have outriggers help resist typhoons during and after construction, the shim plate correction method described earlier was applied to outrigger-to-belt connection details. Horizontal shims fill gaps between belt truss pockets as shown in Figure 3.26 and outrigger ends.

Figure 3.27: 300 North LaSalle, Chicago. © Marshall Gerometta/CTBUH

▲ Figure 3.28: 300 North LaSalle – core-and-outrigger system © Magnusson Klemencic Associates

▲ Figure 3.29: 300 North LaSalle – outrigger trusses. © Thornton Tomasetti

300 North LaSalle
Chicago, USA

This 57-story building completed in 2009 has a concrete core, steel perimeter columns and floor framing, and steel belt and outrigger trusses at Levels 40–41 (see Figures 3.28 & 3.29). The outrigger location corresponds to a mechanical floor located at approximately two-thirds of the building height to serve floors above and below and is visible in the building massing (see Figure 3.27). To reduce outrigger system forces from differential shortening, truss final connections were delayed until the roof slab was cast. Long-term forces induced in outriggers, belts, and columns from core creep were included in the ultimate capacity checks by the engineer.

Chicago Spire (Proposal)
Chicago, USA

The Spire, a 610-meter-tall residential tower in Chicago, had a proposed structure suited the unique architectural vision of the tower (see Figure 3.30). The floor plan, core, and perimeter column grid are based on circles of radii varying with height. In this arrangement the outriggers rely on circular tension and compression rings rather than the usual rectilinear geometry. To minimize visual encumbrance of the perimeter, steel perimeter columns, and steel floor framing were chosen to work with a central reinforced concrete core. Steel outrigger trusses and perimeter steel belt walls at Levels 40, 74, 111, and 140 coincide with transitions in core wall geometry and perimeter column plan locations (see Figures 3.31–3.33). Redistribution of both gravity and lateral loads between the central core and perimeter structure thus occurs at the same locations. The Spire illustrates perhaps the most significant advantage to outrigger systems – almost complete freedom of architectural expression for the tower exterior form.

▲ Figure 3.30: Chicago Spire © Shelbourne
Development / Santiago Calatrava

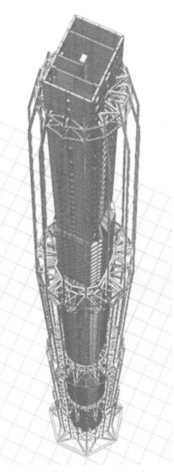

▲ Figure 3.31: Chicago Spire – core and
outriggers. © Thornton Tomasetti

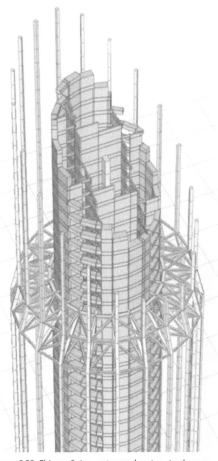

▲ Figure 3.32: Chicago Spire – cutaway showing circular core-
and-outrigger arrangement. © Thornton Tomasetti

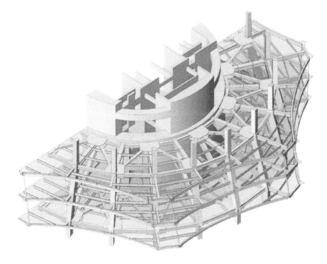

▲ Figure 3.33: Chicago Spire – isometric showing horizontally trussed floors for column kink thrust.
© Thornton Tomasetti

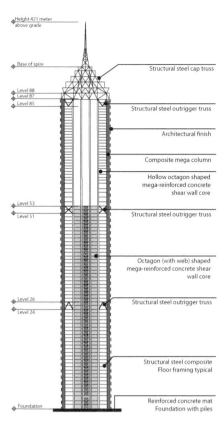

▲ Figure 3.34: Miglin-Beitler Tower, Chicago proposal. © Cesar Pelli Associates

▲ Figure 3.35: Jin Mao, Shanghai. © SOM

▲ Figure 3.36: Jin Mao – structural system elevation. (Source: Korista et al. 1995)

3.5 Ultra Tall Building Outrigger Systems

Miglin-Beitler Skyneedle (Proposal) Chicago, USA

From a historical standpoint, the proposal in the late 1980s for the Miglin-Beitler Skyneedle in Chicago (see Figure 3.34) and the realization of the Jin Mao Tower in Shanghai in 1995 (see next example) can be seen as the precursors of a series of ultra-tall building proposals with similar structural systems. A large reinforced concrete core is connected to a small number of composite steel/concrete mega columns extending over the full height of the tower through sets of multi-story-tall structural steel outrigger trusses at several elevations. This has been a recurrent theme for ultra-tall structural designs over the past 15 years.

The 125-story, 609-meter-tall, 43-meter-wide proposal was never built, but laid significant engineering groundwork for designs to come. The tower was proposed to reside on a very small site and would therefore require extreme slenderness. The architectural design concept did not follow a continuous exterior form, but would instead hearken back to classical designs from the 1930s in New York, which included significant articulation of the perimeter envelope.

Jin Mao Building Shanghai, China

The core-and-outrigger system reached its tacit maturity in the application to the 88-story, 421-meter Jin Mao Building in Shanghai completed in 1999 (see Figure 3.35). The challenge of marrying a highly articulated exterior form with an efficient structural solution in a typhoon-prone region was met by this standard-setting design. An octagonal-shaped reinforced concrete core was linked to eight perimeter composite steel/concrete mega columns which taper and set back to create the unique architectural profile (see Figures 3.36 & 3.37). Steel floor framing members, steel perimeter secondary columns, and composite metal floor decking and concrete slabs complete the mixed design of steel and concrete structural elements (Korista et al. 1995).

The two-story steel outrigger trusses pass through and are encased in the reinforced concrete core walls (see Figure 3.38). To address the issue of the short-term transfer of force through the outrigger trusses, the trusses were erected as the construction progressed, but diagonal to chord connections initially consisted of large diameter pins in slotted holes to create a temporary mechanism and minimize force transfer between core and perimeter columns (see Figure 3.39). Late in construction, once potential future relative movement of the core and perimeter was minimized, the truss connections were "locked up" by placing and tightening connection bolts in the steel diagonal and chord assemblies (see Figure 3.40).

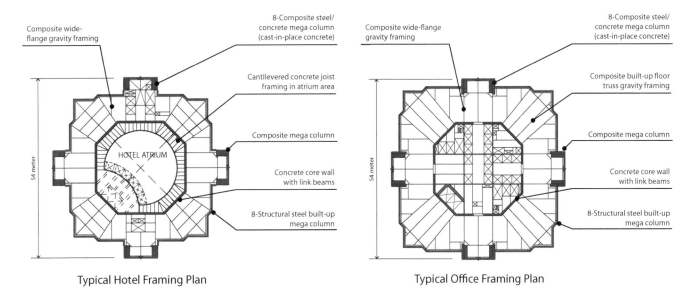

Typical Hotel Framing Plan

Typical Office Framing Plan

▲ Figure 3.37: Jin Mao – typical framing plan. (Source: Korista et al. 1995)

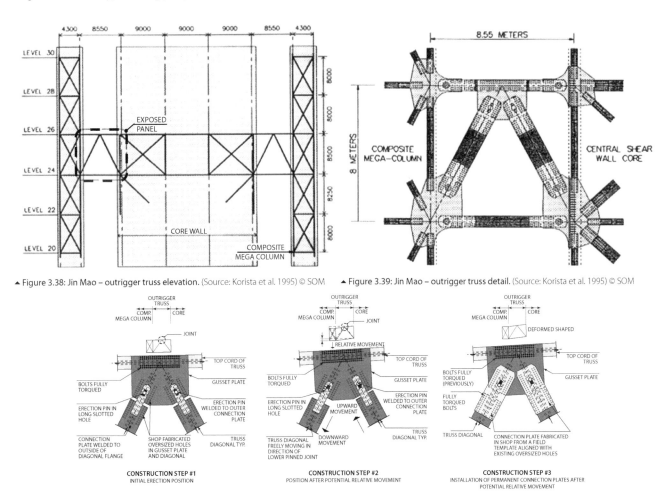

▲ Figure 3.38: Jin Mao – outrigger truss elevation. (Source: Korista et al. 1995) © SOM

▲ Figure 3.39: Jin Mao – outrigger truss detail. (Source: Korista et al. 1995) © SOM

▲ Figure 3.40: Jin Mao – movement at outrigger connection during construction. (Source: Korista et al. 1995) © SOM

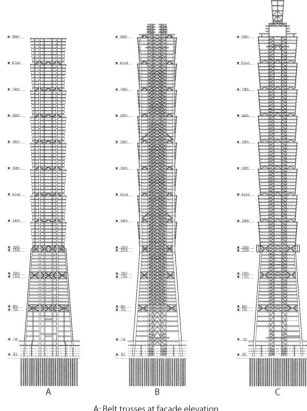

A: Belt trusses at façade elevation
B: Mega columns and outriggers at outer braced core face
C: Outriggers at inner braced core line

▲ Figure 3.41: Taipei 101 showing eight tapering modules of eight stories each above a pyramidal base. © Taipei Financial Center Corporation

▲ Figure 3.42: Taipei 101 – bracing elevations. © Taipei Financial Center Corporation

Taipei 101
Taiwan

Reflecting cultural references and construction preferences, the Taipei 101 in Taiwan completed in 2004, recalls jointed bamboo, tiered pagodas, and the "lucky" number eight. The total height of 508 meters includes a pyramidal base truncated at Level 26 topped by eight, 8-story modules that flare wider with height to create a series of setback floors (see Figure 3.41). This made internal (core) schemes more practical than exterior tube schemes. Potentially severe seismic excitation and a strong steel construction industry favored lightweight steel construction. However, extreme typhoon winds made efficient lateral stiffness a priority. The local building code required refuge areas on multiple floors which were combined with mechanical spaces.

The core is compact thanks to double-deck elevators. The structural design is based on a square core of 16 steel box columns linked by four bracing lines in each direction. Overturning stiffness is enhanced by outrigger sets at 11 levels, with eight lines of steel truss outriggers in each set. At upper floors outer trusses engage eight large steel box columns aligned with core corners, inner trusses indirectly engage the columns through belt trusses that also transfer gravity loads at each setback. From Level 26 down, the inner outriggers engage eight additional box columns. To economically improve stiffness, the core-and-outrigger steel box columns are filled with 10,000 psi (69 MPa) concrete from foundation to Level 62. Perimeter moment frames and interior moment-connected beams have reduced beam section or "dogbone" details at regions of greatest flexural rotation demand to locate ductile hinges away from column faces (see Figure 3.42). The braced core and ductile moment frames form a dual system to address seismic safety (Poon et al. 2002; Joseph et al. 2006).

▲ Figure 3.43: Two International Finance Center (IFC2), Hong Kong, China.
© Antony Wood/CTBUH

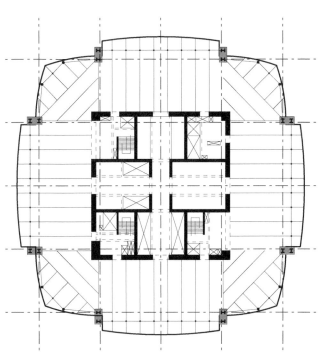

▲ Figure 3.44: IFC2 – typical floor plan. © Arup

Two International Finance Centre Hong Kong, China

This 412-meter-tall tower was completed in 2004. To provide flexible office floor configurations for the tenants connected to the financial industry in this 88-story tower, only eight main mega columns are located in the floor plate with a maximum clear span of 24 meters between columns (see Figures 3.43 & 3.44). Small secondary columns are provided at the corners to control floor slab deflections and vibrations. Following Cheung Kong Centre, this building provides another significant step in the development of core-and-outrigger systems for supertall towers (Luong et al. 2004; Wong 1996).

Three levels of triple-story-high outriggers are provided and located in a straight alignment with the core wall edges. Belt trusses corresponding to the outrigger locations also serve to transfer loads from corner secondary columns to mega columns.

In traditional construction, core walls can advance at three to four days per floor, or two floors per week. Outrigger floors require much more time to allow for the lifting, welding, and installation of special, heavy outrigger system components. Avoiding core wall construction delays would speed the overall schedule. With retro-casting procedures and a very detailed construction plan,

construction speed was as fast as for a normal core wall without outriggers. To do this, blockouts were formed in core walls at outrigger levels.

Outriggers were assembled outside the walls and rolled into place. After outrigger installation the core was backfilled with concrete using retro-casting techniques to form a monolithic element integrating trusses with walls. This approach made construction speed of core walls independent from the outriggers.

Composite columns are common for supertall buildings in China to minimize columns that are already large for strength and stiffness. Embedded continuous steel cores in mega columns and core walls also provide a clear load transfer path for outriggers through steel-to-steel connections.

▲ Figure 3.45: Shanghai Tower. © Gensler

Shanghai Tower
China

The elegant swirling skin of the 126-story, 632-meter Shanghai Tower (see Figure 3.45) scheduled for completion in 2014 conceals a simpler structural frame of stacked cylindrical modules 15 to 17 stories tall with a nine-cell concrete core of roughly 30 square meters, eight main outrigger super columns, four partial-height corner columns, and small secondary columns between them. The modules are separated by two-story refuge and mechanical spaces that include radial steel trusses to support enlarged floors reaching the outer skin and, at eight levels, outriggers with four lines of steel trusses aligned with the core inner "web" walls for maximum separation

of super columns within the circular footprint.

Steel truss outriggers are connected directly to structural steel core members within mega columns and core walls. Composite columns are common for supertall buildings in China to minimize columns that are already large for strength and stiffness. Embedded continuous steel cores in mega columns and core walls also provide a clear load transfer path for outriggers through steel-to-steel connections (see Figures 3.46 & 3.47).

The China code requires design checks under unreduced seismic loading for frequent (50-year), moderate (475-year), and severe (2,475-year) events using

appropriate acceptance criteria. A factor of 1.3 is applied to 50-year "working" loads for strength checks using linear response spectrum analysis. Although there is no explicit strength requirement for the 475-year event, important lateral system structural members are checked using linear response spectrum analysis when requested by the expert review panel. For the 2,475-year event nonlinear analysis is used. A nonlinear time history analysis using scaled seismic event records, required for supertall buildings, revealed that core link beams undergo significant yielding while outrigger trusses neither buckle nor yield. This demonstrates that a core-and-outrigger system can provide ductile performance (Poon et al. 2011).

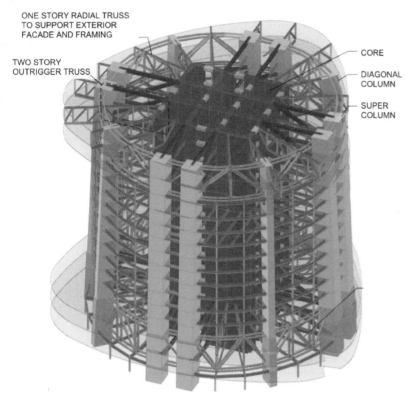

ONE STORY RADIAL TRUSS
TO SUPPORT EXTERIOR
FACADE AND FRAMING

TWO STORY
OUTRIGGER TRUSS

CORE

DIAGONAL
COLUMN

SUPER
COLUMN

▲ Figure 3.46: I Shanghai Tower – isometric of core, mega columns, outrigger, and belt trusses. © Thornton Tomasetti

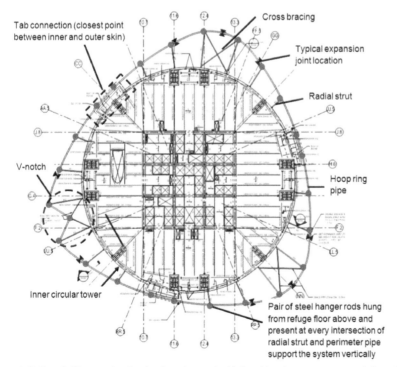

Tab connection (closest point
between inner and outer skin)

Cross bracing

Typical expansion
joint location

Radial strut

V-notch

Hoop ring
pipe

Inner circular tower

Pair of steel hanger rods hung
from refuge floor above and
present at every intersection of
radial strut and perimeter pipe
support the system vertically

▲ Figure 3.47: Shanghai Tower – plan showing floor plate, embedded steel in columns, outer suspended curtain wall support pipe system. © Thornton Tomasetti

▲ Figure 3.48: Plaza Rakyat Office Tower, Kuala Lumpur.
© SOM

3.6 Virtual or Indirect Outrigger Systems (after Nair 1998)

Plaza Rakyat Office Tower Kuala Lumpur, Malaysia

Structural engineers are constantly trying to improve the most thorny and disadvantageous aspects of tall building structural systems. Concerns about slow construction progress at the outrigger levels, and large, relatively unpredictable force redistribution through outrigger elements have concerned structural engineers since the advent of the system 40 years ago. A solution was developed for the partially-constructed Plaza Rakyat 77-story office tower in Malaysia begun in 1998 (see Figure 3.48): the so-called "virtual" or "indirect" outrigger system.

The system involves no direct link between the core and the perimeter frame. Instead it relies upon the perimeter belt wall and the in-plane stiffness and strength of floor slabs linking the core and the perimeter to restrain rotation of the core wall at the "virtual" outrigger levels as discussed earlier in this document. The system was proposed as result of an exhaustive review of the low wind and seismic environmental criteria for the project. The engineers report a significant reduction in lateral drift under wind and fundamental period compared to the core-alone system performance (see Figure 3.49). Designing a virtual outrigger system requires very careful modeling of the floor structures and perimeter belt walls as continuous reinforced concrete elements and not as theoretically rigid diaphragms (see Figure 3.50).

Tower Palace Three Seoul, South Korea

As with the Plaza Rakyat Tower indirect outriggers were used in response to a developing architectural program and the environmental demands of the Tower Palace Three residential project in Seoul, completed in 2004 (see Figure 3.51). When originally planned for 93 stories the structural design included a system of outrigger trusses linked

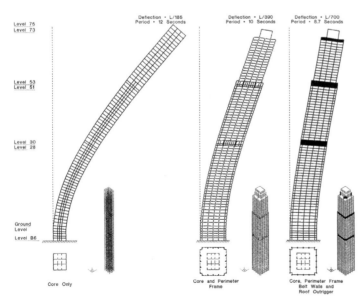

▲ Figure 3.49: Plaza Rakyat – structural diagram. (Source: Viswanath et al. 1998) © SOM

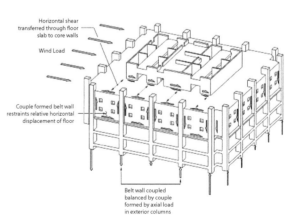

▲ Figure 3.50: Plaza Rakyat – indirect outrigger behavior through slabs and walls to perimeter columns. (Source: Viswanath et al. 1998) © SOM

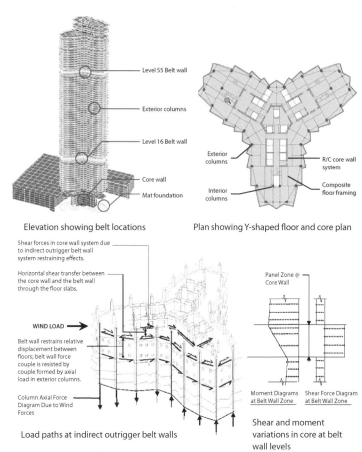

Level 55 Belt wall

Exterior columns

Level 16 Belt wall

Core wall

Mat foundation

Elevation showing belt locations

Exterior columns

Interior columns

R/C core wall system

Composite floor framing

Plan showing Y-shaped floor and core plan

Shear forces in core wall system due to indirect outrigger belt wall system restraining effects.

Horizontal shear transfer between the core wall and the belt wall through the floor slabs.

Panel Zone @ Core Wall

WIND LOAD →

Belt wall restrains relative displacement between floors; belt wall force couple is resisted by couple formed by axial load in exterior columns.

Column Axial Force Diagram Due to Wind Forces

Moment Diagrams at Belt Wall Zone

Shear Force Diagram at Belt Wall Zone

Load paths at indirect outrigger belt walls

Shear and moment variations in core at belt wall levels

to a central triangular core. When the tower was reduced to 69 stories and 273 meters above grade, rather than reconfigure the entire system, the design team proposed simply deleting the direct outrigger trusses and relying on perimeter belt walls already in the original design (Abdelrazaq et al. 2004; Abdelrazaq et al. 2005).

Fundamental structural action of the indirect outrigger system is depicted in Figure 3.52. Indirect outriggers at mechanical levels on Levels 16 and 55 use 0.8 meter thick belt walls and 0.3 meter thick floor slabs. Advantages of indirect outriggers include less restriction on equipment layouts, belt

wall installation off the critical path with no significant delays on construction progress, elimination of force transfers through outrigger elements, and reduction in the extensive detailing and reinforcement coordination required for outriggers and their connections at the core and perimeter frame.

3.7 Mechanically Damped Outrigger Systems

St. Francis Shangri-La Place
Manila, Philippines

Outrigger damping was incorporated in the design of the two residential towers of the St Francis Shangri-La Place

completed in 2009, each 217 meter in height (see Figure 3.53), located within two kilometers of an active seismic fault and subject to typhoons. Each tower has pairs of viscous dampers connecting outriggers to columns. The dampers act when relative movement occurs between outrigger tips and outrigger columns, so this system provides a significant increase in damping but a smaller increase in overturning stiffness than would be provided by a traditional stiffly-connected outrigger system.

Mechanically damped outrigger systems were not discussed within general design guidelines as they represent untraditional outrigger

▲ Figure 3.53: St. Francis Shangri-La Place, Manila. © Mike Gonzalez

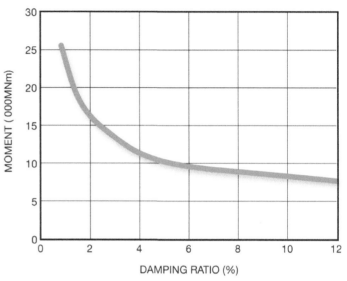

▲ Figure 3.54: Overturning Moment vs. Damping Ratio. (Source: Smith & Willford 2008) © Arup

applications. However they are certainly relevant for tall slender buildings which are frequently sensitive to crosswind excitation from vortex-induced oscillations (VIO) that can adversely affect occupant comfort and generate large overturning forces in windy conditions. The dampers at Shangri-La Place are intended to reduce building acceleration by 35% of the original value with a damping ratio of 7.5% of critical damping. In addition to improving service level wind response, the strength design level wind overturning moments are reduced by 40%.

Wind behavior control is typically an important criterion for tall building design, and often has a major influence on the structural design. One way to reduce VIO effects is through building shape modifications, such as the double-stairstep corners on Taipei 101 that disrupt vortex formation. Another approach is to alter building dynamic properties, by changing building mass or stiffness, but that can be expensive or impractical.

A third approach to improve occupant comfort is supplementary damping, which can be efficient and cost-effective. Damping is well understood and widely accepted by the engineering community for mitigating dynamic load effects. For a hypothetical 400-meter flexible tower with minimal inherent damping levels, supplementary damping could reduce the dynamic overturning moment by approximately a factor of three (see Figure 3.54). As a practical matter the reduction in force is seldom taken for strength design purposes, because of difficulty in guaranteeing that the damping will be in place during extreme wind events; the damping device may be subject to deactivation for maintenance or repair.

Supplementary damping can take the form of viscous dampers, viscoelastic dampers, tuned mass dampers (TMDs), tuned liquid column dampers, or sloshing dampers. Viscous dampers work at all frequencies, generate greater resistance as the driving velocity increases, and convert motion to heat based on the resistance times travel distance. Such dampers are most efficient, compact, and cost-effective when driven through large travel distances at high velocities. While outriggers typically serve as rigid connectors between a core and perimeter columns to increase stiffness and strength against overturning, the geometric leverage offered by outriggers can also be used to drive supplementary mechanical damping devices: large relative movements between outrigger tips and perimeter columns can efficiently drive relatively compact dampers bridging between them.

In contrast to the outrigger approach, TMDs drive dampers through relative movements between the building and a heavy swinging mass, raising the question of tuning for maximum effectiveness: TMDs are tuned to a particular frequency and could become "untuned" and inefficient during and after the extreme events anticipated for strength design, because building frequency changes occur as structures reach strength-level strains. Mechanically damped outriggers can provide similar supplementary damping contributions without the space, weight, or tuning requirements of a TMD.

Modern viscous dampers can be designed for a nonlinear response to driving velocities. The Shangri-La Place dampers were optimized for the ultimate wind condition and damper performance was carefully assessed

for wind behavior subject to constant cyclic loading (see Figures 3.55 & 3.56). The potential for high energy dissipation for a short period during a seismic event was also considered. For example, damper piston velocity is up to 10 mm/s in wind loading and up to 200 mm/s under seismic loading. Since damper force is a function of velocity as well as piston area, fluid viscosity, and orifice size, the high velocity under seismic loading could potentially generate very large resistance. Dampers were designed to limit the resistance force through a pressure release valve. Even if this valve were to fail, the outriggers have been designed to yield in a ductile manner but remain intact.

While dampers are typically used only to reduce building accelerations at service level wind loads, at the Shangri-La Place towers the redundant design of the viscous damper system permitted building ultimate loads to be based on damped behavior as well. Whenever relying on dampers to resist ultimate wind loads, consider providing more dampers than required for "optimum" performance, using dampers acting in parallel, separating dampers from each

other, designing for several dampers failing, assuming the damper system is not behaving 100% efficiently, and designing the structure such that with the failure of all dampers the building will not collapse although damage may be sustained. If these items are accounted for in the design, the contribution of the dampers for reduction of the ultimate or strength-level loads can be incorporated (Korista et al. 1995).

For application of mechanically damped outriggers on future projects, perhaps a formalized probabilistic approach to strength design, such as that used in aircraft design for safety, would be helpful.

Using supplementary damping to decrease wind response permits reductions in required structural stiffness and associated material and labor costs. Smaller column sizes result in an increase in net floor area. Dampers on outriggers also avoid accumulation of outrigger forces from differential shortening. These savings can more than offset the additional costs for the dampers, testing, and installation of this system.

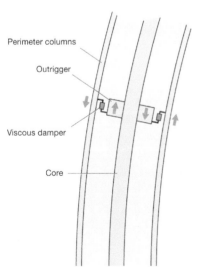

▲ Figure 3.55: Damped outrigger concept. (Source: Smith & Willford 2008) © Arup

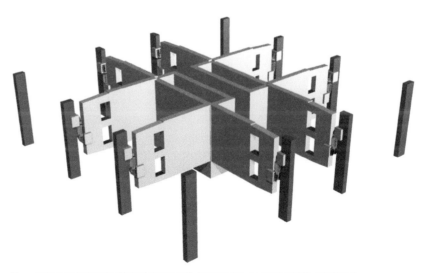

▲ Figure 3.56: Isometric view of lateral system with dampers. (Source: Smith & Willford 2008) © Arup

4.0 Recommendations and Future Research

4.0 Recommendations and Future Research

4.1 Recommendations

Building core-and-outrigger systems have been used for half a century, but have kept evolving to reflect changes in preferred materials, building proportions, analysis methods, and design approaches. The great efficiencies provided by high-strength, high-stiffness concrete, and outriggers connected to mega columns gathering all perimeter loads, have made outrigger systems desirable for tall, slender towers. Variations on the theme, including steel and mixed designs, belt trusses distributing outrigger loads among multiple smaller columns, and indirect or "virtual" outriggers, have found application through a wide range of heights.

While practical implementation of outrigger systems still requires considerable thought, care, and project-specific studies, common conditions and behaviors can be summarized in design guidelines as follows.

Appropriate Conditions for Outrigger Systems:

- Relatively slender lateral systems.

- Lateral drift behavior is cantilever flexure (tall core), not shear (moment frames).

- Able to resize perimeter columns for required stiffness rather than just strength.

Load Transfer Paths in Outrigger Systems:

- Core story shear reversal at outriggers.

- Core story shear magnitude at outriggers may be larger than at typical floors.

- Connect direct outriggers for push-pull on core at primary shear planes.

- Detail to distribute push-pull forces along the core width.

- Connect direct outriggers for vertical forces at core and at outrigger columns.

- Direct outriggers may experience unintended gravity force transfers; see *"Differential Column Shortening Effects"* below.

- Connect indirect or "virtual" outriggers for horizontal diaphragm shear forces to core and belt truss at each chord level.

- Indirect or "virtual"outriggers may avoid unintended gravity force transfers to core.

- Belts as used for indirect outriggers may experience unintended gravity force transfers among perimeter columns due to differential shortening.

- Belts can serve as transfer trusses where some columns are re-spaced or deleted.

- Relying on the same belt for lateral stiffness and gravity transfers may not be appropriate in high seismic areas if belt member yielding is anticipated.

Determining Location of Outriggers in Elevation:

- Ideally, multiple outriggers are distributed along building height, but the benefit from each added outrigger must be weighed against added construction cost and time.

- Direct or conventional outriggers are typically located at mechanical and/or refuge floors.

- Consider superdiagonal outriggers through occupied space subject to owner approval.

- Consider indirect, "virtual" outriggers where direct outrigger diagonals are not acceptable.

- Optimizing outrigger locations and material quantity distribution among core, outriggers, and outrigger columns is an indeterminate process because all parameters are related.

- Column sizes can change while keeping overall axial stiffness between outrigger levels.

- Perform multiple analysis iterations if net tension on reinforced or composite concrete columns leads to reduced, cracked-section properties under some load combinations.

Diaphragm Floors:

- Release diaphragm from direct outrigger chords to avoid "losing" outrigger truss chord axial forces or overstating outrigger truss stiffness in a rigid-diaphragm model.

- Check direct outrigger chords resolve the horizontal component of diagonal members.

- Model realistic diaphragm stiffness at outrigger levels and for at least several stories above and below the outrigger levels to avoid unrealistic, impractical induced column moments and shears and sudden, unrealistic shifts of core wall forces.

- Indirect or "virtual" outriggers require high diaphragm strength and stiffness through thicker, more heavily reinforced slabs or underslab bracing systems.

- Bracket modeled diaphragm stiffness to check for sensitivity of results. Gross uncracked properties will be unrealistically high, transformed area of slab rebar will be unrealistically low, 25% of uncracked properties may be a reasonable starting point.

- Appropriate stiffness may differ for strength-level "ultimate" wind or seismic loads, service-level "working" wind loads and frequent-storm "occupant comfort" wind acceleration checks.

Stiffness Reduction:

- Perform multiple analysis iterations if net tension on reinforced or composite concrete columns leads to reduced, cracked-section properties under some load combinations.

- Consider lateral stiffness reduction from geometric nonlinearity (the P-Delta Effect) either explicitly within the analysis software or as a separate stiffness reduction factor.

- Bracket diaphragm stiffness properties considering the forces acting at that time.

Differential Column Shortening Effects:

- Direct outriggers equalize vertical shortening of core and connected perimeter columns.

- Large unintended force transfers can occur where separate core and column shortening would be different due to gravity loads, creep, or shrinkage.

- Timing of shortening can differ due to material properties, member surface/volume ratios, or exposure to drying conditions, leading to time-related force transfers.

- Predictions of long-term shortening for nonlinear time-dependent materials such as concrete require careful creep and shrinkage testing of planned concrete mixtures.

- Minimize force transfers by designing for similar shortening at core-and-outrigger columns through member sizing, reinforcing, or material properties.

- Minimize force transfers at dissimilar vertical shortening by delaying final outrigger connections as late as possible in construction. Only later movements induce forces.

- Use "staged construction" analysis models to predict pre- and post-connection differential shortening movements and the resulting effects on the structure.

- Temporary jack and shim connections can release transfer forces before fixing. Linked jacks have also been proposed.

- Include force transfers in design load combinations. Vary load factor depending on loads included in the combination and probability of simultaneous occurrence.

- Force transfers between core and perimeter columns are avoided at indirect or "virtual" outriggers but stiff belt trusses enforce shortening compatibility among adjacent perimeter columns which could result in some unintended vertical force transfers there.

Thermal Effects Management:

- Load combination factors on forces from simultaneous thermal strain, differential shortening, gravity loads, and lateral loads should reflect realistic probabilities.

- Thermal strains should consider exposure to uncontrolled temperature changes, including radiant heating and cooling (solar gain and black box radiation) and heat flow along members to and from conditioned spaces.

Load Path from Connections:

- Similar materials are easiest to connect. Transferring forces between different materials such as steel outrigger truss members to concrete core walls is more difficult.

The great efficiencies provided by high-strength, high-stiffness concrete, and outriggers connected to mega columns gathering all perimeter loads, have made outrigger systems desirable for tall, slender towers.

- Consider behavior of connection devices. For example, developing full strength of an embedded headed shear stud requires local slippage and deformation. If such movement is not acceptable, use studs at less than full capacity.

- Consider force distribution within members through embedded columns or stubs, embedded chord reinforcing or embedded truss members.

Panel Zone Load Path:

- Core shears and reversals at outriggers are analogous to moment frame panel zones.

Taking a broader view, a research program cannot "solve" outrigger design and construction challenges since they include a variety of situations and solutions, with new concepts being developed. Construction will always reflect location-specific preferences, abilities and limitations.

- Check core walls or bracing at these locations. Minimize wall penetrations there.

- Check coupling or link beams for large axial compression or tension forces from outrigger chords plus large shear forces from panel zone behavior.

- Consider strut-and-tie design approach where opening layouts permit.

- Where reinforced concrete alone is insufficient, consider if embedded steel trusses can be used to reinforce concrete core walls against high "panel zone" shear at outrigger levels.

Outrigger System Construction Sequence:

- Delay direct outrigger connections to reduce inadvertent gravity load transfer between core and columns from differential shortening. Jacks and shims can provide outrigger behavior where required before final lock-in.

- Non-typical construction at outriggers delays progress. Delay reduction strategies include separate erection of outrigger trusses, then sliding them into wall pockets.

- Local requirements influence construction strategies. Sometimes outrigger connections can be delayed until top-out, while in other situations outriggers must be in service during construction to resist typhoons and permit early move-in on lower floors.

Code Interpretations for Seismic Load Resisting Systems:

- Outrigger systems are not specifically listed in model building codes.

- Typical seismic philosophy of ductile post-yield behavior in a large event is not appropriate for outriggers.

- Performance Based Design can demonstrate acceptable behavior of outrigger systems by showing the outriggers remain elastic even under large seismic events.

Soft-Story and Weak-Story Seismic Requirements:

- Soft-story and weak-story seismic requirements should not be applicable in relation to the stiff stories inherent in an outrigger system.

Strong Column Seismic Requirement and Capacity Based Design:

- The strong column seismic requirement calls for columns designed to remain ductile in an event, or capable of resisting amplified seismic design forces, or shown to not be overloaded by actual forces through Performance Based Design. Only the PBD approach is appropriate for most outrigger systems.

- Controlled capacity of buckling restrained brace outriggers can limit load on columns.

Strong Column Weak Beam Concept in Outrigger Systems:

- The strong column weak beam seismic requirement should not be applicable to the outrigger column and the outrigger truss. The core alone provides ample stiffness and strength against story collapse from column hinging.

Capacity Based Connection Design:

▸ Relatively massive outrigger members for effective stiffness cannot have connections to be stronger than member capacities, a conventional seismic design approach.

▸ Performance Based Design can demonstrate that lesser, practical connections will still remain substantially elastic under the largest seismic event required by code.

Mechanically Damped Outrigger Systems:

▸ Where supplementary damping is found critical to tower performance, outriggers have been used as rigid arms to amplify building movements and drive viscous dampers.

4.2 Future Research

The building examples provided in this document illustrate a wide range of design approaches and construction strategies successfully applied to tall building outrigger systems over half a century. Each strategy has its own set of pros and cons, so there is ample opportunity for exercising engineering creativity and judgment for years to come.

Rather than representing a specific discipline or body of research, outriggers reflect numerous aspects of structural behavior. Two examples are differential shortening prediction and diaphragm stiffness modeling. Currently designers must select one (or more) formulation for creep and shrinkage and incorporate it into a staged construction analysis, with no ability to evaluate its predictive ability in the future. And designers must "bracket" possible diaphragm behaviors for modeling stiffness in their analyses. Certainly better guidance on these topics, along with many others, would be helpful.

Taking a broader view, a research program cannot "solve" outrigger design and construction challenges since they include a variety of situations and solutions, with new concepts being developed. Construction will always reflect location-specific preferences, abilities, and limitations. For future research, the greatest value of this guideline document is to stimulate thoughtful discussions of outrigger systems within the engineering profession, encourage researchers to investigate behaviors related to such systems, and address and resolve many of the issues through subsequent editions. The authors look forward to participating with the rest of the tall building community in these exciting developments.

5.0 References

Bibliography

Allard, F. & Santamouris, M. (1998) *Natural Ventilation in Buildings: A Design Handbook.* Routledge: New York.

Abdelrazaq, A., Baker, W., Chung, K., Pawlikowski, J., Wang, I. & Yon, K. (2004) "Integration of Design and Construction of the Tallest Building in Korea, Tower Palace III, Seoul, Korea." *Proceedings of CTBUH 2004 Seoul Conference – Tall Buildings in Historical Cities – Culture & Technology for Sustainable Cities,* CTBUH: Chicago, pp. 654–661.

Abdelrazaq, A., Kijewski-Correa, T., Song, Y., Case, P., Isyumov, N. & Kareem, A. (2005) "Design and Full-scale Monitoring of the Tallest Building in Korea: Tower Palace III." *Proceedings of 6th Asia-Pacific Conference on Wind Engineering.* Techno Press: Seoul.

Ali, M. and Kyoung, S. (2007) "Structural Developments in Tall Buildings: Current Trends and Future Prospects." *Architectural Science Review,* Vol. 50.3, pp. 205–223.

Arbitrio, V. & Chen, K. (2005) "300 Madison Avenue Practical Defensive Design Meets Post 9/11 Challenge." *Structure Magazine,* April, pp. 35–36.

American Institute of Steel Construction (AISC). (2002) *AISC 341-02: Seismic Provisions for Structural Steel Buildings.* AISC: Chicago.

American Society of Civil Engineers (ASCE). (2005) *ASCE 7-05: Minimum Design Loads for Buildings and Other Structures.* ASCE: Reston, USA.

American Society of Civil Engineers (ASCE). (2010) *ASCE 7-10: Minimum Design Loads for Buildings and Other Structures.* ASCE: Reston, USA.

Baker, W., James, P., Tomlinson, R. & Weiss, A. (2009) "Trump International Hotel & Tower." *CTBUH Journal 2009 Issue III,* pp. 16–22

Baker, W., Korista, D., Novak, L., Pawlikowski, J. & Young, B. (2007) "Creep and Shrinkage and the Design of Supertall Buildings – A Case Study: The Burj Dubai Tower." *American Concrete Institute Symposium Publications,* SP-246, pp. 133–148.

Baker, W., Korista, S., Sinn, R., Pennings, K. & Rankin, D. (2006) "Trump International Hotel and Tower." *Concrete International,* July 2006, pp. 28–32.

Bayati, Z., Mahdikhani, M. & Rahaei, A. (2008) "Optimized Use of Multi-outriggers System to Stiffen Tall Buildings." *Proceedings of 14th World Conference on Earthquake Engineering.* International Association of Earthquake Engineering (IAEE): Tokyo.

Beedle, L. & Iyengar, H. (1982) "Selected Works of Fazlur R. Khan (1929–1982). IABSE Structure C-23/82.

Callow, J., Krall, K. & Scarangello, T. (2009) "Inside Out." *Modern Steel Construction,* January 2009, pp. 21–25.

Chen, K. & Axmann, G. (2003) "Comprehensive Design and A913 Grade 65 Steel Shapes: the Key Design Factors of 300 Madison Avenue, New York City." *2003 NASCC Proceedings, Sessions D20/C26.* NASCC: Baltimore, pp. 1–9.

Cheng, S., Liu, J., Jin, Z. & Bao, Z. (1998) "A Model Shaking Table Test for Shanghai Ciro's Plaza (in Chinese)." *Building Science,* Vol. 14 (5), pp. 8–13.

Chung, K., Scott, D., Kim, D., Ha, I. & Park, K. (2008) "Structural System of North East Asia Trade Tower in Korea."

Proceedings of CTBUH 8th World Congress. Council on Tall Buildings and Urban Habitat: Chicago, pp. 425–432.

Derrick, R. & Rodrigues, N. (2008) "Design of the Tallest Reinforced Concrete Structure in California – A 58-story Residential Tower in San Francisco." *Proceedings of the 2008 Structures Congress: Crossing Borders.* ASCE: Reston, USA, pp. 1–9.

Gerasimidis, S., Efthymiou, E. & Baniotopoulos, C. (2009) "Optimum Outrigger Locations of High-rise Steel Buildings for Wind Loading." *Proceedings of 5th European-African Conference on Wind Engineering (EACWE).* International Association for Wind Engineering (IAWE): Tokyo.

Joseph, L., Poon, D. & Shieh, S. (2006) "Ingredients of High-Rise Design: Taipei 101, the World's Tallest Building." *Structure Magazine,* June 2006, pp. 40–45.

Kareem, A., Kijewski, T. & Tamura, Y. (1999) "Mitigation of Motions of Tall Buildings with Specific Examples of Recent Applications." *Wind and Structures,* Vol. 2, No. 3, pp. 201–251.

Khan, Y. (2004) *Engineering Architecture: The Vision of Fazlur R. Khan.* W.W. Norton & Co.: New York, pp. 176–183.

Korista, S., Sarkisian, M. & Abdelrazaq, A. (1995) "Jin Mao Tower's Unique Structural System." Paper presented at the 1995 Shanghai International Seminar for Building Construction Technology ('95 SISBCT).

Kwok, M. & Vesey, D. (1997) "Reaching for the Moon – A view on the Future of Tall Buildings." *Structures in the New Millennium, Proceedings of the Fourth International Kerensky Conference.* A.A. Balkema: Amsterdam, pp.199–205.

Lahey, J., Wolf, M., Klemencic, R. & Johansson, O. (2008) "A Tale of Two Cities: Collaborative Innovations for Sustainable Towers." *Proceedings of CTBUH 8th World Congress.* Council on Tall Buildings and Urban Habitat: Chicago, pp. 362–372.

Lame, A. (2008) "Optimization of Outrigger Structures." MEng. Thesis. Massachusetts Institute of Technology, June 2008.

Loesch, E. (2007) "An Enduring Solution: WaMu Center/Seattle Art Museum Expansion." *Structure Magazine,* June 2007, pp. 46–48.

Luong, A., Gibbons, C., Lee, A. & MacArthur, J. (2004) "Two International Finance Centre." *Proceedings of CTBUH 2004 Seoul Conference – Tall Buildings in Historical Cities – Culture & Technology for Sustainable Cities,* CTBUH: Chicago. pp. 1,160–1,164.

Moehle, J. (2007) "The Tall Buildings Initiative for Alternative Seismic Design." *The Structural Design of Tall and Special Buildings,* Vol. 16, Issue 5, pp. 559–567.

Nair, R. (1998) "Belt Trusses and Basements as 'Virtual' Outriggers for Tall Buildings." *Engineering Journal,* Fourth Quarter, pp. 140–146.

Nolte, C. (2006) "Tall, Skinny…Stable. Using Novel Technology, S.F. Tower Should Resist Quakes, Gales." *San Francisco Chronicle,* July 2, 2006, pp. B-1.

Po, S. & Siahaan, F. (2001) "The Use of Outrigger and Belt Truss System for High-rise Concrete Buildings." *Dimensi Teknik Sipil,* vol. 3, no. 1, March, pp. 36–40.

Poon, D., Hsiao, L., Zhu, Y., Joseph, L., Zuo, S., Fu, P. & Ihtiyar, O. (2011) "Non-Linear Time History Analysis for the Performance Based Design of

Shanghai Tower." *Proceedings of the 2011 Structures Congress.* ASCE: Reston, USA, pp. 541–551.

Poon, D., Shieh, S., Joseph, L. & Chang, C. (2002) "The Sky's the Limit." *Modern Steel Construction,* December 2002, pp. 24–28.

Scarangello, T., Krall, K. & Callow, J. (2008) "A Statement in Steel: The New York Times Building." *Proceedings of CTBUH 8th World Congress.* Council on Tall Buildings and Urban Habitat: Chicago, pp. 654–659

Shahrooz, B., Tunc, G. & Deason, J. (2004) "Outrigger Beam-Wall Connections II: Subassembly Testing and Further Modeling Enhancements." *Journal of Structural Engineering,* Vol. 130, Issue 2, pp. 262–270.

Steel Institute of New York (SINY). (2006) "The New York Times Building: Steel Structures Disappear into the Sky." *METALS in Construction,* Fall 2006, pp. 20–27.

Smith, B. & Coull, A. (2007) *Tall Building Structures Analysis and Design.* John Wiley & Sons: New Jersey.

Smith, R. & Willford, M. (2008) "Damped Outriggers for Tall Buildings." *The Arup Journal,* 3/2008, pp. 15–21.

Taranath, B. (1988) *Structural Analysis & Design of Tall Buildings.* McGraw Hill: Texas.

Taranath, B. (1998) *Steel, Concrete and Composite Design of Tall Buildings.* McGraw Hill: New York.

Taranath, B. (2011) *Structural Analysis and Design of Tall Buildings: Steel and Composite Construction.* CRC Press: Boca Raton.

Tomasetti, T., Poon, D. & Hsiao, L. (2001) "The Tallest Concrete Building in Shanghai, China – Plaza 66." *Proceedings of the 6th World Congress of the CTBUH – Tall Buildings and Urban Habitat – Cities in the Third Millennium.* Spon Press: London, pp. 719–727.

Viswanath, H., Tolloczko, J. & Clarke, J. (eds.) (1998) *Multi-Purpose High-Rise Towers and Tall Buildings.* E & FN Spon: London, pp. 333–346.

Wada, A. (1990) "How to Reduce Drift of Buildings." *ATC-15-3 Proceedings of Fourth US-Japan Workshop on the Improvement of Building Structural Design and Construction Practice.* Applied Technology Council: Redwood City, USA, pp. 349–365.

Willford, M., Whittaker, A. & Klemencic, R. (2008) *Recommendations for the Seismic Design of High-rise Buildings. A Consensus Document – CTBUH Seismic Working Group.* Council on Tall Buildings and Urban Habitat: Chicago.

Willford, M. & Smith, R. (2008) "Performance Based Seismic and Wind Engineering for 60-story Twin Towers in Manila." *Proceedings of 14th World Conference on Earthquake Engineering.* International Association of Earthquake Engineering (IAEE): Tokyo.

Wong, R. (1996) "Construction of Two IFC." Accessed April 2012. http://bst1.cityu.edu.hk/e-learning/building_info_pack/tall_building/ifc2_const.pdf.

Youssef, N., Wilkerson, R., Fischer, K. & Tunick, D. (2010) "Seismic Performance of a 55-storey Steel Plate Shear Wall." *The Structural Design of Tall and Special Buildings,* Vol. 19, Issue 1–2, pp. 139–165.

The Council on Tall Buildings and Urban Habitat is the official arbiter of the criteria upon which tall building height is measured, and the title of "The World's (or Country's, or City's) Tallest Building" determined. The Council maintains an extensive set of definitions and criteria for measuring and classifying tall buildings which are the basis for the official "100 Tallest Buildings in the World" list (see pages 83–86).

What is a Tall Building?

There is no absolute definition of what constitutes a "tall building." It is a building that exhibits some element of "tallness" in one or more of the following categories:

- **Height relative to context:** It is not just about height, but about the context in which it exists. Thus, whereas a 14-story building may not be considered a tall building in a high-rise city such as Chicago or Hong Kong, in a provincial European city or a suburb this may be distinctly taller than the urban norm.
- **Proportion:** Again, a tall building is not just about height but also about proportion. There are numerous buildings which are not particularly high, but are slender enough to give the appearance of a tall building, especially against low urban backgrounds. Conversely, there are numerous big/large footprint buildings which are quite tall but their size/floor area rules them out as being classed as a tall building.
- **Tall Building Technologies:** If a building contains technologies which may be attributed as being a product of "tall" (e.g., specific vertical transport technologies, structural wind bracing as a product of height, etc.), then this building can be classed as a tall building.

Although number of floors is a poor indicator of defining a tall building due to the changing floor-to-floor height between differing buildings and functions (e.g., office versus residential usage), a building of perhaps 14 or more stories—or over 50 meters (165 feet) in height—could perhaps be used as a threshold for considering it a "tall building."

What are Supertall and Megatall Buildings?

The CTBUH defines "supertall" as a building over 300 meters (984 feet) in height, and a "megatall" as a building over 600 meters (1,968 feet) in height. Although great heights are now being achieved with built tall buildings—in excess of 800 meters (2,600 feet)—as of July 2012 there are only approximately 65 supertall and 2 megatall buildings completed and occupied globally.

How is a tall building measured?

The CTBUH recognizes three categories for measuring building height (see diagrams opposite):

1. **Height to Architectural Top:** Height is measured from the level[1] of the lowest, significant,[2] open-air,[3] pedestrian[4] entrance to the architectural top of the building, including spires, but not including antennae, signage, flagpoles, or other functional-technical equipment.[5] This measurement is the most widely utilized and is employed to define the Council on Tall Buildings and Urban Habitat (CTBUH) rankings of the "World's Tallest Buildings."

2. **Highest Occupied Floor:** Height is measured from the level[1] of the lowest, significant,[2] open-air,[3] pedestrian[4] entrance to the finished floor level of the highest occupied[6] floor within the building.

3. **Height to Tip:** Height is measured from the level[1] of the lowest, significant,[2] open-air,[3] pedestrian[4] entrance to the highest point of the building, irrespective of material or function of the highest element (i.e., including antennae, flagpoles, signage, and other functional-technical equipment).

Number of Floors:

The number of floors should include the ground floor level and be the number of main floors above ground, including any significant mezzanine floors and major mechanical plant floors. Mechanical mezzanines should not be included if they have a significantly smaller floor area than the major floors below. Similarly, mechanical penthouses or plant rooms protruding above the general roof area should not be counted. Note: CTBUH floor counts may differ from published accounts, as it is common in some regions of the world for certain floor levels not to be included (e.g., the level 4, 14, 24, etc. in Hong Kong).

Building Usage:
What is the difference between a tall building and a telecommunications/observation tower?

A tall "building" can be classed as such (as opposed to a telecommunications/observation tower) and is eligible for the "tallest" lists if at least 50 percent of its height is occupied by usable floor area.

Single-Function and Mixed-Use Buildings:

- A **single-function** tall building is defined as one where 85 percent or more of its total floor area is dedicated to a single usage.
- A **mixed-use** tall building contains two or more functions (or uses), where each of the functions occupy a significant proportion[7] of the tower's total space. Support areas such as car parks and mechanical plant space do not

World's tallest 10 buildings according to "Architectural Height" *(as of February 2014)*

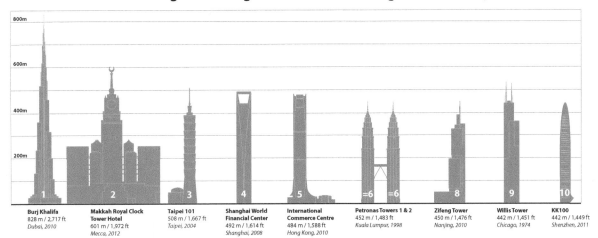

Burj Khalifa
828 m / 2,717 ft
Dubai, 2010

Makkah Royal Clock Tower Hotel
601 m / 1,972 ft
Mecca, 2012

Taipei 101
508 m / 1,667 ft
Taipei, 2004

Shanghai World Financial Center
492 m / 1,614 ft
Shanghai, 2008

International Commerce Centre
484 m / 1,588 ft
Hong Kong, 2010

Petronas Towers 1 & 2
452 m / 1,483 ft
Kuala Lumpur, 1998

Zifeng Tower
450 m / 1,476 ft
Nanjing, 2010

Willis Tower
442 m / 1,451 ft
Chicago, 1974

KK100
442 m / 1,449 ft
Shenzhen, 2011

World's tallest 10 buildings according to "Highest Occupied Floor" *(as of February 2014)*

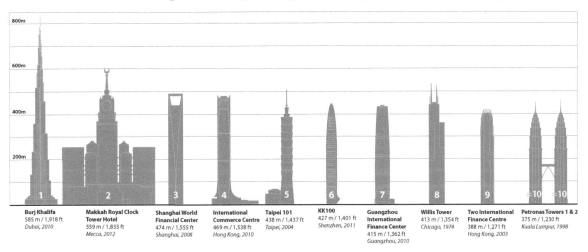

Burj Khalifa
585 m / 1,918 ft
Dubai, 2010

Makkah Royal Clock Tower Hotel
559 m / 1,833 ft
Mecca, 2012

Shanghai World Financial Center
474 m / 1,555 ft
Shanghai, 2008

International Commerce Centre
469 m / 1,538 ft
Hong Kong, 2010

Taipei 101
438 m / 1,437 ft
Taipei, 2004

KK100
427 m / 1,401 ft
Shenzhen, 2011

Guangzhou International Finance Center
415 m / 1,362 ft
Guangzhou, 2010

Willis Tower
413 m / 1,354 ft
Chicago, 1974

Two International Finance Centre
388 m / 1,271 ft
Hong Kong, 2003

Petronas Towers 1 & 2
375 m / 1,230 ft
Kuala Lumpur, 1998

World's tallest 10 buildings according to "Height to Tip" *(as of February 2014)*

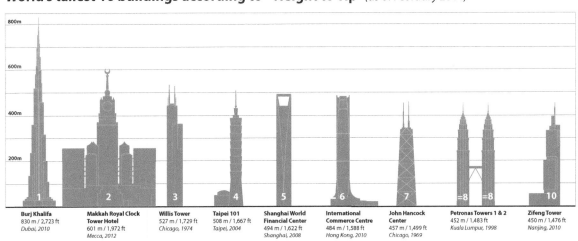

Burj Khalifa
830 m / 2,723 ft
Dubai, 2010

Makkah Royal Clock Tower Hotel
601 m / 1,972 ft
Mecca, 2012

Willis Tower
527 m / 1,729 ft
Chicago, 1974

Taipei 101
508 m / 1,667 ft
Taipei, 2004

Shanghai World Financial Center
494 m / 1,622 ft
Shanghai, 2008

International Commerce Centre
484 m / 1,588 ft
Hong Kong, 2010

John Hancock Center
457 m / 1,499 ft
Chicago, 1969

Petronas Towers 1 & 2
452 m / 1,483 ft
Kuala Lumpur, 1998

Zifeng Tower
450 m / 1,476 ft
Nanjing, 2010

constitute mixed-use functions. Functions are denoted on CTBUH "tallest" lists in descending order, e.g., "hotel/office" indicates hotel above office function.

Building Status:

- **Complete (Completion):**
  A building is considered to be "complete" (and added to the CTBUH Tallest Buildings lists) if it fulfills all of the following three criteria: (i) topped out structurally and architecturally, (ii) fully clad, and (iii) open for business, or at least occupiable.

- **Under Construction (Start of Construction):**
  A building is considered to be "under construction" once site clearing has been completed and foundation/piling work has begun.

- **Topped Out:**
  A building is considered to be "topped out" when it is under construction, and has reached its full height both structurally and architecturally (e.g., including its spires, parapets, etc.).

- **Proposed (Proposal):**
  A building is considered to be "proposed" (i.e., a real proposal) when it fulfills all of the following criteria: (i) has a specific site with ownership interests within the building development team, (ii) has a full professional design team progressing the design beyond the conceptual stage, (iii) Has obtained, or is in the process of obtaining, formal planning consent/legal permission for construction, and (iv) has a full intention to progress the building to construction and completion.

- **Vision:**
  A building is considered to be a "vision" when it either: (i) is in the early stages of inception and does not yet fulfill the criteria under the "proposal" category, or (ii) was a

proposal that never advanced to the construction stages, or (iii) was a theoretical proposition.

- **Demolished:**
  A building is considered to be "demolished" after it has been destroyed by controlled end-of-life demolition, fire, natural catastrophe, war, terrorist attack, or through other means intended or unintended.

Structural Material:

- A **steel** tall building is defined as one where the main vertical and lateral structural elements and floor systems are constructed from steel.
- A **concrete** tall building is defined as one where the main vertical and lateral structural elements and floor systems are constructed from concrete.
- A **composite** tall building utilizes a combination of both steel and concrete acting compositely in the main structural elements, thus including a steel building with a concrete core.
- A **mixed-structure** tall building is any building that utilizes distinct steel or concrete systems above or below each other. There are two main types of mixed structural systems: a **steel/concrete** tall building indicates a steel structural system located above a concrete structural system, with the opposite true of a **concrete/steel** building

Additional Notes on Structure:

(i) If a tall building is of steel construction with a floor system of concrete planks on steel beams, it is considered a **steel** tall building.
(ii) If a tall building is of steel construction with a floor system of a concrete slab on steel beams, it is considered a **steel** tall building.
(iii) If a tall building has steel columns plus a floor system of concrete beams, it is considered a **composite** tall building.

Footnotes:

[1] Level: finished floor level at threshold of the lowest entrance door.

[2] Significant: the entrance should be predominantly above existing or pre-existing grade and permit access to one or more primary uses in the building via elevators, as opposed to ground floor retail or other uses which solely relate/connect to the immediately adjacent external environment. Thus entrances via below-grade sunken plazas or similar are not generally recognized. Also note that access to car park and/or ancillary/support areas are not considered significant entrances.

[3] Open-air: the entrance must be located directly off of an external space at that level that is open to air.

[4] Pedestrian: refers to common building users or occupants and is intended to exclude service, ancillary, or similar areas.

[5] Functional-technical equipment: this is intended to recognize that functional-technical equipment is subject to removal/addition/change as per prevalent technologies, as is often seen in tall buildings (e.g., antennae, signage, wind turbines, etc. are periodically added, shortened, lengthened, removed and/or replaced).

[6] Highest occupied floor: this is intended to recognize conditioned space which is designed to be safely and legally occupied by residents, workers or other building users on a consistent basis. It does not include service or mechanical areas which experience occasional maintenance access, etc.

[7] This "significant proportion" can be judged as 15 percent or greater of either: (i) the total floor area, or (ii) the total building height, in terms of number of floors occupied for the function. However, care should be taken in the case of supertall towers. For example a 20-story hotel function as part of a 150-story tower does not comply with the 15 percent rule, though this would clearly constitute mixed-use.

100 Tallest Buildings in the World (as of February 2014)

The Council maintains the official list of the 100 Tallest Buildings in the World, which are ranked based on the height to architectural top, and includes not only completed buildings, but also buildings currently under construction. However, a building does not receive an official ranking number until it is completed (see criteria, pages 80–82).

Color Key:
Buildings listed in black are completed and officially ranked.
Buildings listed in green are under construction and have topped out.
Buildings listed in red are under construction, but have not yet topped out.

Rank	Building Name	City	Year	Stories	Height m	Height ft	Material	Use
	Kingdom Tower	Jeddah	2019	167	1000 **	3281	concrete	residential / hotel / office
1	Burj Khalifa	Dubai	2010	163	828	2717	steel / concrete	office / residential / hotel
	Suzhou Zhongnan Center	Suzhou	–	138	700 **	2297	–	residential / hotel / office
	Ping An Finance Center	Shenzhen	2016	115	660	2165	composite	office
	Wuhan Greenland Center	Wuhan	2017	125	636	2087	composite	hotel / residential / office
	Shanghai Tower	Shanghai	2015	128	632	2073	composite	hotel / office
2	Makkah Royal Clock Tower Hotel	Mecca	2012	120	601	1972	steel / concrete	other / hotel / multiple
	Goldin Finance 117	Tianjin	2016	128	597	1957	composite	hotel / office
	Lotte World Tower	Seoul	2016	123	555	1819	composite	hotel / office
	One World Trade Center	New York City	2014	94	541	1776	composite	office
	The CTF Guangzhou	Guangzhou	2017	111	530	1739	composite	hotel / residential / office
	Tianjin Chow Tai Fook Binhai Center	Tianjin	2017	97	530	1739	composite	residential / hotel / office
	Zhongguo Zun	Beijing	2018	108	528	1732	composite	office
3	Taipei 101	Taipei	2004	101	508	1667	composite	office
4	Shanghai World Financial Center	Shanghai	2008	101	492	1614	composite	hotel / office
5	International Commerce Centre	Hong Kong	2010	108	484	1588	composite	hotel / office
	International Commerce Center 1	Chongqing	2017	99	468	1535	composite	hotel / office
	Guangdong Building	Tianjin	2017	91	468	1535	composite	residential / hotel / office
	Lakhta Center	St. Petersburg	2018	86	463	1517	composite	office
	Riverview Plaza A1	Wuhan	2016	82	460	1509	–	hotel / office
	The Wharf IFS	Suzhou	2016	92	452	1483	composite	residential / hotel / office
	Changsha IFS Tower T1	Changsha	2016	88	452	1483	composite	residential / office
6	Petronas Tower 1	Kuala Lumpur	1998	88	452	1483	composite	office
6	Petronas Tower 2	Kuala Lumpur	1998	88	452	1483	composite	office
8	Zifeng Tower	Nanjing	2010	66	450	1476	composite	hotel / office
9	Willis Tower	Chicago	1974	108	442	1451	steel	office
	World One	Mumbai	2015	117	442	1450	composite	residential
10	KK100	Shenzhen	2011	100	442	1449	composite	hotel / office
11	Guangzhou International Finance Center	Guangzhou	2010	103	439	1439	composite	hotel / office
	Wuhan Center	Wuhan	2015	88	438	1437	composite	hotel / residential / office
	Dream Dubai Marina	Dubai	2014	101	432	1417	concrete	serviced apartments / hotel
	Diamond Tower	Jeddah	2017	93	432	1417	–	residential
	432 Park Avenue	New York City	2015	85	426	1397	concrete	residential
12	Trump International Hotel & Tower	Chicago	2009	98	423	1389	concrete	residential / hotel
13	Jin Mao Tower	Shanghai	1999	88	421	1380	composite	hotel / office
14	Princess Tower	Dubai	2012	101	413	1356	steel / concrete	residential
15	Al Hamra Tower	Kuwait City	2011	80	413	1354	concrete	office
16	Two International Finance Centre	Hong Kong	2003	88	412	1352	composite	office
	LCT Landmark Tower	Busan	2018	101	412	1350	–	hotel / residential
	Huaguoyuan Tower 1	Guiyang	2017	64	406	1332	composite	–
	Huaguoyuan Tower 2	Guiyang	2017	64	406	1332	composite	–
	Nanjing Olympic Suning Tower	Nanjing	2017	88	400	1312	steel / concrete	residential / hotel / office
	China Resources Headquarters	Shenzhen	2017	–	400	1312	–	office
17	23 Marina	Dubai	2012	90	393	1289	concrete	residential
18	CITIC Plaza	Guangzhou	1996	80	390	1280	concrete	office
	Logan Century Center 1	Nanning	2017	82	386	1266	–	hotel / office
	Capital Market Authority Headquarters	Riyadh	2014	77	385	1263	composite	office
19	Shun Hing Square	Shenzhen	1996	69	384	1260	composite	office
	Eton Place Dalian Tower 1	Dalian	2014	80	383	1257	composite	hotel / office
	Abu Dhabi Plaza	Astana	2017	88	382	1253	–	residential

* estimated height
** minimum height

Rank	Building Name	City	Year	Stories	Height m	ft	Material	Use
	World Trade Center Abu Dhabi - The Residences	Abu Dhabi	2014	88	381	1251	concrete	residential
20	Empire State Building	New York City	1931	102	381	1250	steel	office
21	Elite Residence	Dubai	2012	87	380	1248	concrete	residential
22	Central Plaza	Hong Kong	1992	78	374	1227	concrete	office
	Oberoi Oasis Tower B	Mumbai	2016	82	372	1220	concrete	residential
	The Address The BLVD	Dubai	2016	72	370	1214	concrete	residential / hotel
	Golden Eagle Tiandi Tower A	Nanjing	–	76	368	1208	–	hotel / office
23	Bank of China Tower	Hong Kong	1990	72	367	1205	composite	office
24	Bank of America Tower	New York City	2009	55	366	1200	composite	office
	Dalian International Trade Center	Dalian	2015	86	365	1199	composite	residential / office
	VietinBank Business Center Office Tower	Hanoi	2017	68	363	1191	composite	office
	Federation Towers - Vostok Tower	Moscow	2016	93	360	1181	concrete	residential / hotel / office
25	Almas Tower	Dubai	2008	68	360	1181	concrete	office
25	The Pinnacle	Guangzhou	2012	60	360	1181	concrete	office
27	JW Marriott Marquis Hotel Dubai Tower 1	Dubai	2012	82	355	1166	concrete	hotel
27	JW Marriott Marquis Hotel Dubai Tower 2	Dubai	2013	82	355	1166	concrete	hotel
29	Emirates Tower One	Dubai	2000	54	355	1163	composite	office
	Oko Tower 1	Moscow	2015	85	352	1155	concrete	residential / hotel
	Forum 66 Tower 2	Shenyang	2015	68	351	1150	composite	office
	Hanking Center	Shenzhen	2018	65	350	1148	–	office
	J97	Changsha	2014	97	349	1146	steel	residential / office
30	Tuntex Sky Tower	Kaohsiung	1997	85	348	1140	composite	hotel / office
31	Aon Center	Chicago	1973	83	346	1136	steel	office
32	The Center	Hong Kong	1998	73	346	1135	steel	office
33	John Hancock Center	Chicago	1969	100	344	1128	steel	residential / office
	Four Seasons Place	Kuala Lumpur	2017	65	343	1124	–	residential / hotel
	ADNOC Headquarters	Abu Dhabi	2014	76	342	1122	concrete	office
	Ahmed Abdul Rahim Al Attar Tower	Dubai	2014	76	342	1122	steel / concrete	residential
	Xiamen International Centre	Xiamen	2016	61	340	1115	composite	office
	LCT Residential Tower A	Busan	2018	85	339	1113	–	residential
	The Wharf Times Square 1	Wuxi	2015	68	339	1112	composite	hotel / residential
	Chongqing World Financial Center	Chongqing	2014	73	339	1112	composite	office
34	Mercury City Tower	Moscow	2013	75	339	1112	concrete	residential / office
	Tianjin Modern City	Tianjin	2015	65	338	1109	composite	residential / hotel
	Orchid Crown Tower A	Mumbai	2016	75	337	1106	concrete	residential
	Orchid Crown Tower B	Mumbai	2016	75	337	1106	concrete	residential
35	Tianjin World Financial Center	Tianjin	2011	75	337	1105	composite	office
36	The Torch	Dubai	2011	79	337	1105	concrete	residential
37	Keangnam Hanoi Landmark Tower	Hanoi	2012	72	336	1102	concrete	hotel / residential / office
	Wilshire Grand Tower	Los Angeles	2017	73	335	1100	steel / concrete	hotel / office
	DAMAC Residenze	Dubai	2016	86	335	1099	steel / concrete	residential
38	Shimao International Plaza	Shanghai	2006	60	333	1094	concrete	hotel / office
	LCT Residential Tower B	Busan	2018	85	333	1093	–	residential
	Mandarin Oriental Hotel	Chengdu	2017	88	333	1093	–	residential / hotel
39	Rose Rayhaan by Rotana	Dubai	2007	71	333	1093	composite	hotel
	China Chuneng Tower	Shenzhen	2016	–	333	1093	–	–
40	Modern Media Center	Changzhou	2013	57	332	1089	composite	office
41	Minsheng Bank Building	Wuhan	2008	68	331	1086	steel	office
	Ryugyong Hotel	Pyongyang	–	105	330	1083	concrete	hotel / office
	Gate of Kuwait Tower	Kuwait City	2016	84	330	1083	concrete	hotel / office
42	China World Tower	Beijing	2010	74	330	1083	composite	hotel / office
	Thamrin Nine Tower 1	Jakarta	–	71	330	1083	–	office
	Zhuhai St. Regis Hotel & Office Tower	Zhuhai	–	67	330	1083	–	hotel / office
	The Skyscraper	Dubai	–	66	330	1083	–	office
	Yuexiu Fortune Center Tower 1	Wuhan	2016	66	330	1083	composite	office
	Suning Plaza Tower 1	Zhenjiang	2016	77	330	1082	composite	–
	Hon Kwok City Center	Shenzhen	2015	80	329	1081	composite	residential / office
43	Longxi International Hotel	Jiangyin	2011	72	328	1076	composite	residential / hotel
43	Al Yaqoub Tower	Dubai	2013	69	328	1076	concrete	hotel
	Nanjing World Trade Center Tower 1	Nanjing	2016	69	328	1076	composite	hotel / office
	Golden Eagle Tiandi Tower B	Nanjing	–	68	328	1076	–	office
	Wuxi Suning Plaza 1	Wuxi	2014	68	328	1076	composite	hotel / office
	Concord International Centre	Chongqing	2016	62	328	1076	composite	hotel / office
	Greenland Center Tower 1	Qingdao	2016	74	327	1074	composite	hotel / office
45	The Index	Dubai	2010	80	326	1070	concrete	residential / office
	Cemindo Tower	Jakarta	2015	63	325 *	1066	concrete	hotel / office

Rank	Building Name	City	Year	Stories	m	ft	Material	Use
					Height			
46	The Landmark	Abu Dhabi	2013	72	324	1063	concrete	residential / office
46	Deji Plaza	Nanjing	2013	62	324	1063	composite	hotel / office
	Yantai Shimao No. 1 The Harbour	Yantai	2014	59	323	1060	composite	residential / hotel / office
48	Q1 Tower	Gold Coast	2005	78	323	1058	concrete	residential
	Lamar Tower 1	Jeddah	2016	70	322	1056	concrete	residential / office
49	Wenzhou Trade Center	Wenzhou	2011	68	322	1056	concrete	hotel / office
	Guangxi Finance Plaza	Nanning	2016	68	321	1053	composite	hotel / office
50	Burj Al Arab	Dubai	1999	56	321	1053	composite	hotel
51	Nina Tower	Hong Kong	2006	80	320	1051	concrete	hotel / office
	Palais Royale	Mumbai	2014	88	320	1050	concrete	residential
	White Magnolia Plaza 1	Shanghai	2015	66	320	1048	composite	office
	Chongqing IFS T1	Chongqing	2016	64	320	1048	composite	–
52	Chrysler Building	New York City	1930	77	319	1046	steel	office
	Global City Square	Guangzhou	2015	67	319	1046	composite	office
53	New York Times Tower	New York City	2007	52	319	1046	steel	office
	Runhua Global Center 1	Changzhou	2015	72	318	1043	composite	office
	Jiuzhou International Tower	Nanning	2015	71	318	1043	–	–
	Riverside Century Plaza Main Tower	Wuhu	2015	66	318	1043	composite	hotel / office
54	HHHR Tower	Dubai	2010	72	318	1042	concrete	residential
	Yunrun International Tower	Huaiyin	2016	75	317	1040	–	office
55	Bank of America Plaza	Atlanta	1993	55	317	1040	composite	office
	Changsha IFS Tower T2	Changsha	2016	–	315	1033	composite	office
	Youth Olympics Center Tower 1	Nanjing	2015	68	315	1032	composite	–
	Maha Nakhon	Bangkok	2016	77	313	1028	concrete	residential / hotel
	The Stratford Residences	Makati	2015	74	312	1024	concrete	residential
	Moi Center Tower A	Shenyang	2014	75	311	1020	composite	hotel / office
56	U.S. Bank Tower	Los Angeles	1990	73	310	1018	steel	office
57	Ocean Heights	Dubai	2010	83	310	1017	concrete	residential
57	Menara Telekom	Kuala Lumpur	2001	55	310	1017	concrete	office
	Bodi Center Tower 1	Hangzhou	2016	55	310	1017	–	office
	Fortune Center	Guangzhou	2015	73	309	1015	composite	office
59	Pearl River Tower	Guangzhou	2012	71	309	1015	composite	office
60	Emirates Tower Two	Dubai	2000	56	309	1014	concrete	hotel
	Eurasia	Moscow	2014	72	309	1013	composite	hotel / office
	Guangfa Securities Headquarters	Guangzhou	2016	62	308	1010	–	office
	Burj Rafal	Riyadh	2014	68	308	1010	concrete	residential / hotel
	Wanda Plaza 1	Kunming	2016	67	307	1008	composite	office
	Wanda Plaza 2	Kunming	2016	67	307	1008	composite	office
61	Cayan Tower	Dubai	2013	73	307	1008	concrete	residential
	Lokhandwala Minerva	Mumbai	2015	83	307	1007	concrete	residential
62	Franklin Center - North Tower	Chicago	1989	60	307	1007	composite	office
	One57	New York City	2014	79	306	1005	steel / concrete	residential / hotel
63	East Pacific Center Tower A	Shenzhen	2013	85	306	1004	concrete	residential
63	The Shard	London	2013	73	306	1004	composite	residential / hotel / office
65	JPMorgan Chase Tower	Houston	1982	75	305	1002	composite	office
66	Etihad Towers T2	Abu Dhabi	2011	80	305	1002	concrete	residential
67	Northeast Asia Trade Tower	Incheon	2011	68	305	1001	composite	residential / hotel / office
68	Baiyoke Tower II	Bangkok	1997	85	304	997	concrete	hotel
	Wuxi Maoye City - Marriott Hotel	Wuxi	2014	68	304	997	composite	hotel
	Shenzhen World Finance Center	Shenzhen	2016	68	304	997	composite	office
69	Two Prudential Plaza	Chicago	1990	64	303	995	concrete	office
	Diwang International Fortune Center	Liuzhou	2014	75	303	994	composite	residential / hotel / office
	KAFD World Trade Center	Riyadh	2014	67	303	994	concrete	office
	Jiangxi Nanchang Greenland Central Plaza 1	Nanchang	2014	59	303	994	composite	office
	Jiangxi Nanchang Greenland Central Plaza 2	Nanchang	2014	59	303	994	composite	office
70	Leatop Plaza	Guangzhou	2012	64	303	993	composite	office
71	Wells Fargo Plaza	Houston	1983	71	302	992	steel	office
72	Kingdom Centre	Riyadh	2002	41	302	992	steel / concrete	residential / hotel / office
73	The Address	Dubai	2008	63	302	991	concrete	residential / hotel
	Gate of the Orient	Suzhou	2014	68	302	990	composite	residential / hotel / office
74	Capital City Moscow Tower	Moscow	2010	76	302	990	concrete	residential
	Greenland Puli Center	Jinan	2015	61	301	988	composite	residential / office
	Heung Kong Tower	Shenzhen	2014	70	301	987	composite	hotel / office
	Brys Buzz	Greater Noida	2017	82	300	984	concrete	residential

Rank	Building Name	City	Year	Stories	Height m	ft	Material	Use
75	Doosan Haeundae We've the Zenith Tower A	Busan	2011	80	300	984	concrete	residential
	Supernova Spira	Noida	2016	80	300	984	concrete	residential
	NBK Tower	Kuwait City	2016	70	300	984	concrete	office
	Huachuang International Plaza Tower 1	Changsha	2016	66	300	984	composite	hotel / office
	Torre Costanera	Santiago	2014	64	300	984	concrete	office
	Riverfront Times Square	Shenzhen	2016	64	300	984	composite	hotel / office
75	Abeno Harukas	Osaka	2014	62	300	984	steel	hotel / office / retail
	Namaste Tower	Mumbai	2015	62	300 *	984	concrete	hotel / office
	Golden Eagle Tiandi Tower C	Nanjing	–	60	300	984	–	office
75	Arraya Tower	Kuwait City	2009	60	300	984	concrete	office
	Shenglong Global Center	Fuzhou	2016	57	300	984	–	office
75	Aspire Tower	Doha	2007	36	300	984	composite	hotel / office
	Shum Yip Upperhills Tower 2	Shenzhen	–	–	300	984	–	office
	Jin Wan Plaza 1	Tianjin	2015	66	300	984	–	hotel / office
	Langham Hotel Tower	Dalian	2015	74	300	983	–	residential / hotel
79	First Bank Tower	Toronto	1975	72	298	978	steel	office
79	One Island East	Hong Kong	2008	68	298	978	concrete	office
	Ilham Baru Tower	Kuala Lumpur	2015	64	298	978	concrete	residential / office
	Yujiapu Yinglan International Finance Center	Tianjin	2016	60	298	978	composite	office
	Four World Trade Center	New York City	2014	64	298	977	composite	office
81	Eureka Tower	Melbourne	2006	91	297	975	concrete	residential
	Dacheng Financial Business Center Tower A	Kunming	2015	–	297	974	steel	hotel / office
82	Comcast Center	Philadelphia	2008	57	297	974	composite	office
83	Landmark Tower	Yokohama	1993	73	296	972	steel	hotel / office
	R&F Yingkai Square	Guangzhou	2014	66	296	972	composite	residential / hotel / office
84	Emirates Crown	Dubai	2008	63	296	971	concrete	residential
	Xiamen Shimao Cross-Strait Plaza Tower B	Xiamen	2015	67	295	969	–	office
85	Khalid Al Attar Tower 2	Dubai	2011	66	294	965	concrete	hotel
	Lamar Tower 2	Jeddah	2016	62	293	961	concrete	residential / office
86	311 South Wacker Drive	Chicago	1990	65	293	961	concrete	office
87	Sky Tower	Abu Dhabi	2010	74	292	959	concrete	residential / office
88	Haeundae I Park Marina Tower 2	Busan)	2011	72	292	958	composite	residential
89	SEG Plaza	Shenzhen	2000	71	292	957	concrete	office
	Indiabulls Sky Suites	Mumbai	2015	75	291	955	concrete	residential
90	70 Pine Street	New York City	1932	67	290	952	steel	office
	Hunter Douglas International Plaza	Guiyang	2014	69	290	951	composite	hotel / office
	Tanjong Pagar Centre	Singapore	2016	68	290	951	–	residential / hotel / office
	Powerlong Center Tower 1	Tianjin	2015	59	290	951	composite	office
	Zhengzhou Eastern Center North Tower	Zhengzhou	2016	78	289	948	composite	office
	Zhengzhou Eastern Center South Tower	Zhengzhou	2016	78	289	948	composite	office
91	Dongguan TBA Tower	Dongguan	2013	68	289	948	composite	hotel / office
	Busan International Finance Center Landmark Tower	Busan	2014	63	289	948	concrete	office
92	Key Tower	Cleveland	1991	57	289	947	composite	office
93	Shaoxing Shimao Crown Plaza	Shaoxing	2012	60	288	946	composite	hotel / office
94	Plaza 66	Shanghai	2001	66	288	945	concrete	office
95	One Liberty Place	Philadelphia	1987	61	288	945	steel	office
	Kaisa Center	Huizhou	2015	66	288	945	composite	hotel / office
	International Financial Tower	Dongguan	2016	66	288	945	–	hotel / office
	17 IBC	Moscow	2016	65	288	945	–	–
	18 IBC	Moscow	2016	60	288	945	–	office
	Colorful Yunnan City Office Tower	Kunming	2016	59	288	945	–	office
96	Yingli International Finance Centre	Chongqing	2012	58	288	945	concrete	office
	Soochow International Plaza East Tower	Huzhou	2014	50	288	945	composite	hotel / office
	Soochow International Plaza West Tower	Huzhou	2014	50	288	945	composite	residential
97	United International Mansion	Chongqing	2013	67	287	942	concrete	office
98	Chongqing Poly Tower	Chongqing	2013	58	287	941	concrete	office / hotel
	Four Seasons Hotel and Private Residences New York Downtown	New York City	2016	67	286	937	concrete	residential / hotel
99	Millennium Tower	Dubai	2006	59	285	935	concrete	residential
100	Sulafa Tower	Dubai	2010	75	285	935	concrete	residential

CTBUH Organizational Members

(as of February 2014)

Supporting Contributors:
AECOM
Al Hamra Real Estate Company
Broad Sustainable Building Co., Ltd.
Buro Happold, Ltd.
CCDI Group
China State Construction Engineering Corporation (CSCEC)
CITIC Heye Investment (Beijing) Co., Ltd.
Dow Corning Corporation
Emaar Properties, PJSC
Eton Properties (Dalian) Co., Ltd.
Illinois Institute of Technology
Jeddah Economic Company
Kingdom Real Estate Development Co.
Kohn Pedersen Fox Associates, PC
KONE Industrial, Ltd.
Lotte Engineering & Construction Co.
Morin Khuur Tower LLC
National Engineering Bureau
NBBJ
Otis Elevator Company
Renaissance Construction
Samsung C&T Corporation
Shanghai Tower Construction & Development Co., Ltd.
Skidmore, Owings & Merrill LLP
Taipei Financial Center Corp. (TAIPEI 101)
Turner Construction Company
Underwriters Laboratories (UL) LLC
WSP Group

Patrons:
Al Ghurair Construction – Aluminum LLC
Arabtec Construction LLC
Blume Foundation
BMT Fluid Mechanics, Ltd.
The Durst Organization
East China Architectural Design & Research Institute Co., Ltd. (ECADI)
Gensler
Hongkong Land, Ltd.
KLCC Property Holdings Berhad
Langan Engineering & Environmental Services, Inc.
Meinhardt Group International
Permasteelisa Group
Schindler Top Range Division
Shanghai Institute of Architectural Design & Research Co., Ltd.
Studio Daniel Libeskind
Thornton Tomasetti, Inc.
ThyssenKrupp AG
Tishman Speyer Properties
Weidlinger Associates, Inc.
Zuhair Fayez Partnership

Donors:
Adrian Smith + Gordon Gill Architecture, LLP
American Institute of Steel Construction
Aon Fire Protection Engineering Corp.
ARCADIS, US, Inc.
Arup
Aurecon
NV. Besix SA
Brookfield Multiplex Construction Europe Ltd.
C.Y. Lee & Partners Architects/Planners
CH2M HILL
Enclos Corp.
Fender Katsalidis Architects
Halfen USA
Hill International
HOK, Inc.
Jacobs
Laing O'Rourke
Larsen & Toubro, Ltd.
Leslie E. Robertson Associates, RLLP
Magnusson Klemencic Associates, Inc.
MAKE
McNamara / Salvia, Inc.
MulvannyG2 Architecture
Nishkian Menninger Consulting and Structural Engineers
Nobutaka Ashihara Architect PC
PDW Architects
Pei Cobb Freed & Partners
Pickard Chilton Architects, Inc.
PT Gistama Intisemesta
Quadrangle Architects Ltd.
Rafik El-Khoury & Partners
Rolf Jensen & Associates, Inc.
Rowan Williams Davies & Irwin, Inc.
RTKL Associates Inc.
Saudi Binladin Group / ABC Division
Severud Associates Consulting Engineers, PC
Shanghai Construction (Group) General Co. Ltd.
Shree Ram Urban Infrastructure, Ltd.
Sinar Mas Group - APP China
Skanska
Solomon Cordwell Buenz
Studio Gang Architects
SWA Group
Syska Hennessy Group, Inc.
T.Y. Lin International Pte. Ltd.

(List continued on next page)

Tongji Architectural Design (Group) Co., Ltd. (TJAD)
Walter P. Moore and Associates, Inc.
Werner Voss + Partner

Contributors:

Aedas, Ltd.
Akzo Nobel
Allford Hall Monaghan Morris Ltd.
Alvine Engineering
Architectus Brisbane Pty Ltd
Bates Smart
Benoy Limited
Bonacci Group
Boundary Layer Wind Tunnel Laboratory
Bouygues Construction
The British Land Company PLC
Canary Wharf Group, PLC
Canderel Management, Inc.
CBRE Group, Inc.
CCL
Continental Automated Buildings Association (CABA)
CTSR Properties Limited
DBI Design Pty Ltd
DCA Architects Pte Ltd
Deerns Consulting Engineers
DK Infrastructure Pvt. Ltd.
Dong Yang Structural Engineers Co., Ltd.
Far East Aluminium Works Co., Ltd.
GGLO, LLC
Goettsch Partners
Gradient Microclimate Engineering Inc. (GmE)
Graziani + Corazza Architects Inc.
Hariri Pontarini Architects
The Harman Group
Hiranandani Group
Israeli Association of Construction and Infrastructure Engineers (IACIE)
J. J. Pan and Partners, Architects and Planners
Jiang Architects & Engineers
Jones Lang LaSalle Property Consultants Pte Ltd
KHP Konig und Heunisch Planungsgesellschaft
Langdon & Seah Singapore
Lend Lease
Liberty Group Properties
M Moser Associates Ltd.
Mori Building Co., Ltd.
Nabih Youssef & Associates
National Fire Protection Association
National Institute of Standards and Technology (NIST)
National University of Singapore
Norman Disney & Young
OMA Asia (Hong Kong) Ltd.
Omrania & Associates
The Ornamental Metal Institute of New York
Parsons Brinckerhoff
Pei Partnership Architects
Perkins + Will
Philip Chun and Associates Pty Ltd
Pomeroy Studio Pte Ltd
PT Ciputra Property, Tbk
RAW Design Inc.
Ronald Lu & Partners
Royal HaskoningDHV
Sanni, Ojo & Partners
Silvercup Studios
SilverEdge Systems Software, Inc.
Silverstein Properties
SIP Project Managers Pty Ltd
The Steel Institute of New York
Stein Ltd.
Tekla Corporation
Terrell Group
TSNIIEP for Residential and Public Buildings
University of Illinois at Urbana-Champaign
Vetrocare SRL
Wilkinson Eyre Architects
Wirth Research Ltd
Woods Bagot

Participants:

ACSI (Ayling Consulting Services Inc)
Adamson Associates Architects

ADD Inc.
Aidea Philippines, Inc.
AIT Consulting
AKF Group, LLC
AKT II Limited
Al Jazera Consultants
Alimak Hek AB
alinea consulting LLP
Alpha Glass Ltd.
ALT Cladding, Inc.
Altitude Façade Access Consulting
ARC Studio Architecture + Urbanism
ArcelorMittal
Architects 61 Pte., Ltd.
Architectural Design & Research Institute of Tsinghua University
Architectus
Arquitectonica
Atkins
Azrieli Group Ltd.
Azorim Construction Ltd.
Bakkala Consulting Engineers Limited
Baldridge & Associates Structural Engineering, Inc.
BAUM Architects
BDSP Partnership
Beca Group
Benchmark
BG&E Pty., Ltd.
BIAD (Beijing Institute of Architectural Design)
Bigen Africa Services (Pty) Ltd.
Billings Design Associates, Ltd.
bKL Architecture LLC
BluEnt
BOCA Group
Bollinger + Grohmann Ingenieure
Boston Properties, Inc.
Broadway Malyan
Brunkeberg Industriutveckling AB
Buro Ole Scheeren
C S Structural Engineering, Inc.
Callison, LLC
Camara Consultores Arquitectura e Ingeniería
Capital Group
Cardno Haynes Whaley, Inc.
Case Foundation Company
CB Engineers
CCHRB (Chicago Committee on High-Rise Buildings)
CDC Curtain Wall Design & Consulting, Inc.
Central Scientific and Research Institute of Engineering Structures "SRC Construction"
Cermak Peterka Petersen, Inc. (CPP Inc.)
12-Jul-11
China Academy of Building Research
China Institute of Building Standard Design & Research (CIBSDR)
Chinachem Group
City Developments Limited
Concrete Reinforcing Steel Institute (CRSI)
COOKFOX Architects
Cosentini Associates
COWI A/S
Cox Architecture Pty. Ltd.
CS Associates, Inc.
CTL Group
Cubic Architects
Cundall
David Engineers Ltd.
Dar Al-Handasah (Shair & Partners)
Delft University of Technology
Dennis Lau & Ng Chun Man Architects & Engineers (HK), Ltd.
dhk Architects (Pty) Ltd
Diar Consult
DSP Design Associates Pvt., Ltd.
Dunbar & Boardman
Earthquake Engineering Research & Test Center of Guangzhou University
ECSD S.r.l.
Edgett Williams Consulting Group, Inc.
Edmonds International USA
Electra Construction LTD
Elenberg Fraser Pty Ltd
ENAR, Envolventes Arquitectonicas
Ennead Architects LLP

Environmental Systems Design, Inc.
Epstein
Exova Warringtonfire
Farrells
Feilden Clegg Bradley Studios LLP
Fortune Shepler Consulting
FXFOWLE Architects, LLP
Gale International / New Songdo International City Development, LLC
GCAQ Ingenieros Civiles S.A.C.
GEO Global Engineering Consultants
Gilsanz Murray Steficek
M/s. Glass Wall Systems (India) Pvt. Ltd
CCHRB (Chicago Committee on High-Rise Buildings)
Gold Coast City Council
Gorproject (Urban Planning Institute of Residential and Public Buildings)
Grace Construction Products
Gravity Partnership Ltd.
Grimshaw Architects
Grupo Inmobiliario del Parque
Guoshou Yuantong Property Co. Ltd.
GVK Elevator Consulting Services, Inc.
Halvorson and Partners
Handel Architects
Heller Manus Architects
Henning Larsen Architects
Hilson Moran Partnership, Ltd.
Hines
Hong Kong Housing Authority
BSE, The Hong Kong Polytechnic University
Housing and Development Board
IECA Internacional S.A.
ingenhoven architects
Institute BelNIIS, RUE
INTEMAC, SA
Irwinconsult Pty., Ltd.
Ivanhoe Cambridge
Iv-Consult b.v.
Jahn, LLC
Jaros Baum & Bolles
Jaspers-Eyers Architects
JBA Consulting Engineers, Inc.
JCE Structural Engineering Group, Inc.
JMB Realty Corporation
John Portman & Associates, Inc.
Johnson Pilton Walker Pty. Ltd.
Kalpataru Limited
KEO International Consultants
Kinetica
King-Le Chang & Associates
King Saud University College of Architecture & Planning
KPFF Consulting Engineers
KPMB Architects
LBR&A Arquitectos
LCL Builds Corporation
Leigh & Orange, Ltd.
Lerch Bates, Inc.
Lerch Bates, Ltd. Europe
LMN Architects
Lobby Agency
Louie International Structural Engineers
Lyons
Mace Limited
Madeira Valentim & Alem Advogados
MADY
Magellan Development Group, LLC
Margolin Bros. Engineering & Consulting, Ltd.
James McHugh Construction Co.
Meinhardt (Thailand) Ltd.
Metropolis, LLC
Michael Blades & Associates
MKPL Architects Pte Ltd
MMM Group Limited
Moshe Tzur Architects Town Planners Ltd.
MVSA Architects
New World Development Company Limited
Nikken Sekkei, Ltd.
Novawest LLC
NPO SODIS
O'Connor Sutton Cronin
onespace unlimited inc.
Option One International, WLL
Ortiz Leon Arquitectos SLP

P&T Group
Palafox Associates
Paragon International Insurance Brokers Ltd.
Pelli Clarke Pelli Architects
PLP Architecture
Porte Construtora Ltda
PositivEnergy Practice, LLC
Profica Project Management
Project and Design Research Institute "Novosibirsky Promstroyproject"
PT. Prada Tata Internasional (PTI Architects)
Rafael Viñoly Architects, PC
Ramboll
Read Jones Christoffersen Ltd.
Rene Lagos Engineers
RESCON (Residential Construction Council of Ontario)
Rider Levett Bucknall North America
Riggio / Boron, Ltd.
Roosevelt University – Marshall Bennett Institute of Real Estate
Sauerbruch Hutton Gesellschaft von Architekten mbH
schlaich bergermann und partner
Schock USA Inc.
Sematic SPA
Shanghai EFC Building Engineering Consultancy
Shimizu Corporation
Sino-Ocean Land
SKS Associates
Smith and Andersen
SmithGroup
Southern Land Development Co., Ltd.
Sowlat Structural Engineers
Stanley D. Lindsey & Associates, Ltd.
Stauch Vorster Architects
Stephan Reinke Architects, Ltd.
Sufrin Group
Surface Design
Taisei Corporation
Takenaka Corporation
Tameer Holding Investment LLC
Tandem Architects (2001) Co., Ltd.
Taylor Thomson Whitting Pty., Ltd.
TFP Farrells, Ltd.
Thermafiber, Inc.
Transsolar
The Trump Organization
Tyréns
Umow Lai Pty Ltd
University of Maryland – Architecture Library
University of Nottingham
UralNIIProject RAACS
Van Deusen & Associates (VDA)
Vidal Arquitectos
Views On Top Pty Limited
Vipac Engineers & Scientists, Ltd.
VOA Associates, Inc.
Walsh Construction Company
Warnes Associates Co., Ltd
Web Structures Pte Ltd
Werner Sobek Group GmbH
wh-p GmbH Beratende Ingenieure
Windtech Consultants Pty., Ltd.
WOHA Architects Pte., Ltd.
Wong & Ouyang (HK), Ltd.
Wordsearch
WTM Engineers International GmbH
WZMH Architects
Y. A. Yashar Architects
Zemun Ltd.
Ziegler Cooper Architects

Supporting Contributors are those who contribute $10,000; Patrons: $6,000; Donors: $3,000; Contributors: $1,500; Participants: $750.

T - #1034 - 101024 - C88 - 276/216/5 - PB - 9780939493340 - Matt Lamination